BestMasters

More information about this series at http://www.springer.com/series/13198

Springer awards „BestMasters" to the best master's theses which have been completed at renowned universities in Germany, Austria, and Switzerland.

The studies received highest marks and were recommended for publication by supervisors. They address current issues from various fields of research in natural sciences, psychology, technology, and economics.

The series addresses practitioners as well as scientists and, in particular, offers guidance for early stage researchers.

Matthias Würl

Towards Offline PET Monitoring at a Cyclotron-Based Proton Therapy Facility

Experiments and Monte Carlo Simulations

Matthias Würl
Garching, Germany

BestMasters
ISBN 978-3-658-13167-8   ISBN 978-3-658-13168-5   (eBook)
DOI 10.1007/978-3-658-13168-5

Library of Congress Control Number: 2016932889

Springer Spektrum

Printed on acid-free paper

This Springer Spektrum imprint is published by Springer Nature
The registered company is Springer Fachmedien Wiesbaden GmbH

Acknowledgments

This thesis would not have been possible without the help and support of many people. Above all, I would like to show gratitude to my supervisor, Prof. Dr. Katia Parodi, who has supported me throughout this entire project. Japan, Texas, Heidelberg or in the office - no matter how far away, I could always count on immediate and certainly valuable help. Her way of asking questions always made me think twice about things that seemed clear to me. I am glad to have the opportunity to work at her chair for the next few years.

I express my warm thanks to my external supervisor, Dr. Martin Hillbrand, from the Rinecker Proton Therapy Center for his invaluable and kind support and for making this work possible. Hundreds of e-mails and hours of talking provided me a deep insight into clinical practice in proton therapy. It was a true pleasure to perform the PET activation experiments together, although probably many employees of the RPTC wondered about our "early morning exercises".

Furthermore, I would like to thank the team of the RPTC, which was directly or indirectly involved in my research project. Especially Andrea Schieh-Schneider, Želka Geočelović and Cornelia Mühler made the time I had to wait for the PET reconstructions to be ready, appear significantly shorter.

Many problems and questions I had within this project could be answered with the help of scientists from our chair or from HIT. I would like to express gratitude to Dr. Georgios Dedes and Dr. Guillaume Landry for helping me with Monte Carlo and computational issues, as well as Dr. Julia Bauer for her help concerning the PET activation studies. Moreover, I want to thank Franz Englbrecht for the fruitful discussions and the wonderful teamwork - I am looking forward to the next three years...

I would also like to thank the mechanical workshop of our faculty here in Garching, represented by Rolf Oehm for making the phantoms I used in this work.

In the beginning of this project, the PET/CT scanner used for the experiments was a closed book to me. I therefore want to thank Cataldo Tarantino from the Philips customer support and Dr. Christian Zach from Klinikum Großhadern for giving me useful hints and information on this scanner.

Garching would be a really boring place without the presence of my (former) colleagues Eva, Susanne, Ben, Konstantin, Jonas, Michael, Andreas, Daniel, Basti, Tiago, Hugh,

Kathrin, ... Thank you very much for the time we spent here and outside the campus.

Finally, I want to thank all the people that supported me within the last two years and on which I could always count. I want to say thank you to my parents, to my sister Michaela and my brother Thomas, to my friends Tom, Clara, Markus, Jakob, Jonas, Claus, Martin, Tobias, Santiago, Johnny, ... and especially to Jennifer, for always supporting me and trying to understand my work ;-).

Matthias Würl

Abstract

For the full clinical exploitation of the ballistic advantages of proton beams for radiation therapy, accurate knowledge of the location of dose deposition in tissue is necessary. A dedicated tool for monitoring the dose deposition in-situ and in vivo is positron emission tomography (PET). The implementation of this method requires a detailed Monte Carlo model of the clinical proton beamline for the exact dose calculation and β^+-activity prediction. For the latter, reliable cross-section data of the β^+-emitter production channels is needed.

The purpose of this work was therefore to perform two fundamental steps towards the implementation of offline PET monitoring at the Rinecker Proton Therapy Center (RPTC). First, a correct modeling of the proton beam spot scanning and the absolute dose output was established. This included a study of the influence of the low-dose envelope of scanned proton beams on the total dose output. Monoenergetic square fields, as well as homogeneous dose spheres of different sizes were planned with the treatment planning system (TPS) XiO®, and its predicted dose compared to measurements and simulations. Simulated doses agreed nicely with measurements for all square and sphere sizes, but for small squares and spheres the dose predicted by the TPS significantly differs. For treatment plans including small fields it is therefore recommended to verify planned doses by measurements or MC simulations prior to patient irradiation.

The second step towards offline PET monitoring was to validate an experimental cross-section data set at the RPTC for the most relevant β^+-production channels. For this purpose, 3 phantom materials (PE, gelatine and PMMA) were irradiated and the proton-induced activity was measured by a full-ring PET/CT scanner. Production yields were extracted from the measured activities and compared to calculated yields from FLUKA MC simulations and data from HIT and GSI. Integral positron emitter yields measured at the RPTC are comparable to simulated yields using experimental cross-sections. Comparison to simulation results using the internal hadronic model of FLUKA for calculating β^+-emitter yields confirms the findings of previous publications, which recommend to use experimental cross-sections instead. The largest difference to the data from HIT is 18.3 %, whereas yields are heavily overestimated at GSI. The scatter correction applied by the relatively old scanner during PET image reconstruction displaces a significant amount of activity at locations far beyond the proton penetration depth. Therefore, no fine-tuning of the production cross-sections could be performed due to the unreliability of the reconstructed activity spatial distributions for the considered irradiation scenarios.

In order to check the linearity of the PET scanner with the activity level, further irradiations and measurements were performed with different numbers of delivered protons. These data could confirm that due to dead time losses, the scanner underestimates measured activities with increasing level of phantom activation. This resulted in an underestimation of production yields by up to 8 % for the PE phantoms. Instead of delivering as much protons as possible for the sake of good statistics in the dynamically evaluated PET images, a trade-off between high count-rate and dead time losses has to be made when performing thick phantom experiments to validate and fine-tune cross-sections for PET monitoring.

Munich, December 2015

Matthias Würl

Contents

List of Abbreviations

3D-RAMLA	3D Row Action Maximum Likelihood Algorithm
ASCII	American Standard Code for Information Interchange
BGO	bismuth germinate ($Bi_4Ge_3O_{12}$)
BP	Bragg peak
CSDA	Continuous Slowing Down Approximation
CT	Computed Tomography
DICOM	Digital imaging and communication in medicine
FDG	fluorodeoxyglucose (^{18}F)
FLAIR	FLUKA advanced interface
FLUKA	Fluktuierende Kaskade
FOV	field-of-view
FSF	Field Size Factor
FWHM	Full width at half maximum
GSI	Gesellschaft für Schwerionenforschung
GSO	gadolinium oxyorthosilicate (Gd_2SiO_5:Zr)
GUI	Graphical User Interface
HIT	Heidelberger Ionenstrahl-Therapiezentrum
HU	Hounsfield Unit
IC	Ionization Chamber
ICRU	International Commission on Radiation Units and Measurements
IMRT	Intensity-Modulated Radiation Therapy
LOR	line-of-response
LSO	lutetium oxyorthosilicate (Lu_2SiO_5:Ce)
LYSO	lutetium yttrium oxyorthosilicate ($LuYSiO_5$:Ce)
MC	Monte Carlo
MCS	Multiple Coulomb Scattering
MSIC	Multiple Strip Ionization Chamber
MU	Monitor Unit
PE	Polyethylene ($(C_2H_4)_n$)
PET	Positron Emission Tomography
PMMA	Polymethyl methacrylate ($(C_5H_8O_2)_n$)

QA	Quality Assurance
RPTC	Rinecker Proton Therapy Center
SOBP	Spread-out Bragg peak
TIC	Transmission Ionization Chamber
TPS	Treatment Planning System
TRS	Technical Report Series
VMAT	Volumetric Modulated Arc Therapy
WEPL	Water Equivalent Path Length

List of Figures

List of Tables

1 Introduction

The idea to use ionizing radiation for cancer treatment is much older than commonly thought. In 1896, few months after Wilhelm Conrad Röntgen published his discovery of *a new kind of rays* [Röntgen, 1896], L. Freund and E. Schiff suggested to use these rays in the treatment of disease [MacKee, 1921]. From this, biologists and physicians started to investigate the biological effects of ionizing radiation. Physicists developed methods to increase the energy of the x-rays and highly sophisticated delivery techniques. Modern delivery techniques, namely Intensity-Modulated Radiation Therapy (IMRT) and Volumetric Modulated Arc Therapy (VMAT) helped to tremendously reduce the dose delivered to healthy tissue and thus reduce side effects of radiotherapy.

But nevertheless, there is one drawback when using high-energy x-rays for radiation therapy. This is due to the underlying physics of photon interactions in matter and can therefore not be eliminated. Photons get attenuated on their passage through tissue, delivering the largest amount of their energy close to the surface of the tissue. For deep-seated tumors this results in an undesired amount of deposited energy in the healthy tissue around the target volume. One can easily imagine that this is especially undesirable if the tumor is located close to a radiosensitive organ.

To overcome this disadvantage, a different type of radiation is necessary. It was Robert R. Wilson, who in 1946 proposed that accelerated protons could be used for radiotherapy [Wilson, 1946]. The main advantage of ions as compared to photons is their inverse depth-dose distribution. Fast ions deposit less energy to the traversed medium than slow ions, resulting in the so-called Bragg peak at the end of the particle's range. The characteristic depth-dose profile, compared to the profile of high-energy x-rays, can be seen in figure 1.1. Thus, this results in a more accurate dose deposition inside the target volume, while sparing the healthy tissue surrounding the tumor.

But in order to fully exploit these physical advantages, knowledge of the exact location of the energy deposition in tissue is mandatory. If the proton range is not accurately known, severe underdosage in the tumor, leading to bad tumor control, or overdosage in the surrounding healthy tissue, leading to side effects of the therapy, is possible. To overcome this problem, relatively large planning safety margins are applied around the tumor volume [Testa et al., 2009]. One intrinsic reason for these safety margins is that x-ray CT (computed tomography) images are used as patient model for dose planning. This

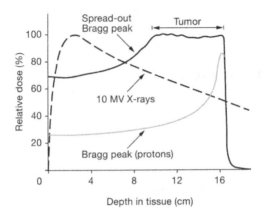

Figure 1.1: Depth-dose profiles of high-energy x-rays and proton beams. In contrast to x-rays (dashed black line), proton beams deposit the largest amount of their energy at the end of their range in the Bragg Peak (gray line). The tumor volume can be covered by superimposing several Bragg curves of different proton beam energies - the spread-out Bragg Peak, SOBP (solid black line). Reprinted by permission from Macmillan Publishers Ltd: Nature Clinical Practice Oncology [Yock and Tarbell, 2004], © 2004.

introduces an uncertainty in ion range of about 1 % to 3 % [Parodi et al., 2007b] due to the unambiguous conversion from CT number to ion stopping power. Furthermore, there are additional sources of uncertainty due to patient positioning and organ motion. However, the large extent of safety margins limits the therapeutic advantage of proton therapy and it would therefore be of general interest to reduce their size. But this can only be done if the ion range is known exactly. It is therefore highly desirable to have a method for ion range verification.

The only clinically implemented method so far is range verification by means of positron emission tomography (PET). Inelastic nuclear collisions of protons with target nuclei can result in the production of β^+-emitters. Fortunately, some of these positron emitters have half-lives in the order of minutes and are thus perfect candidates to be measured with PET scanners during or shortly after irradiation.

There are basically three approaches for treatment verification using PET, an *online* and two *offline* concepts. Online dose verification means to monitor the irradiation-induced β^+-activity during irradiation, or more precisely in the short spill pauses during pulsed beam delivery. This *in-beam* method of course requires that the PET scanner is integrated in the treatment room. At GSI in Darmstadt, this was achieved by a dedicated double head positron camera which was installed directly at the treatment site [Enghardt et al., 2004].

For the two offline concepts, conventional PET or usually combined PET/CT scanners are used to measure the proton-induced activity. The duration of the unavoidable time delay between irradiation and PET acquisition is directly connected to the detectable β^+-emitting isotopes and the count rate. Locating the PET scanner inside the treatment room (*in-room* PET) guarantees high count rates. Hence, only few minutes acquisition time are necessary due to the low time delay between irradiation and measurement. This however prolongs the occupancy time of the treatment room, resulting in lower patient throughput especially for single room facilities.

The other possibility is to measure the activity with a PET/CT scanner in a separate room nearby. The resulting longer time delay between activation and acquisition, which is due to patient transport and repositioning, requires a longer acquisition time of up to 30 min in order to compensate for the reduced activity intensity. This of course affects patient comfort and increases the sensitivity to the degradation of the obtained images due to biological washout processes [Fiedler et al., 2011]. On the other hand, the treatment room is no longer occupied, which potentially allows more patients to be treated. Moreover, an already existing PET/CT scanner acquired for diagnostic purposes can be used for offline PET monitoring. Hence, it is the cheapest way to implement PET monitoring, because no additional installation costs are necessary.

Yet, all these concepts have one thing in common. Since the location of the dose deposition is not identical to the location of the proton-induced β^+-activity, a method that predicts the activity based on the planned dose is needed. For this purpose Monte Carlo (MC) simulations, which are commonly regarded as the computational gold standard for dose calculation in ion beam therapy [Parodi, 2011b], are the method of choice.

But the accuracy of the calculated proton-induced activity distribution strongly depends on the implemented hadronic models of the Monte Carlo code. Several studies indicate that the level of accuracy for activity calculations using the internal models is rather poor for a wide variety of general purpose MC codes [Seravalli et al., 2012, Baumgartl, 2014]. It is therefore recommended to use experimental cross-sections for the MC calculation of positron emitter yields [Parodi et al., 2007a]. However, available cross-section data, e.g. from EXFOR database [Otuka et al., 2014], show large discrepancies between independent data sets [Bauer et al., 2013]. Depending on the considered energy, differences up to several tens of mb can be seen. In figure 1.2, independent data sets for two considered nuclear reaction channels are shown and the non negligible differences between these data sets are obvious. Before using these experimental cross-sections, it is therefore necessary to validate them against measured activation data from the specific PET scanner and irradiated at the specific facility, which shall be used for PET monitoring [Seravalli et al., 2012].

One main ingredient for the implementation of PET monitoring in a proton therapy facility is hence a validated and possibly facility-fine-tuned cross-section data set for the most relevant β^+-production channels. The other important ingredient is an accurate Monte Carlo model of the clinical beamline, where absolute values of the three-dimensional dose distribution in phantoms or patients can be predicted within clinical requirements for any arbitrary treatment field.

The purpose of this study was therefore to establish the MC beam model of the Ri-

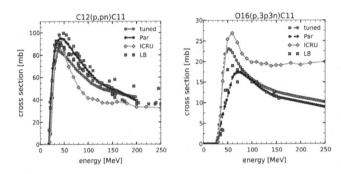

Figure 1.2: Comparison of different cross-section data sets as a function of energy for two of the β^+-emitter production channels studied in this work: ^{12}C(p,pn)^{11}C and ^{16}O(p,3p3n)^{11}C [Bauer et al., 2013]. Their tuned production cross-sections (circles) are compared to the compilation of Parodi et al. [2007b] (hexagons), which is based amongst others on an experimental data series of Iljinov et al. [1991] (squares), and to the theoretical values according to ICRU [2000]. From Bauer et al. [2013], © Institute of Physics and Engineering in Medicine. Reproduced by permission of IOP Publishing. All rights reserved.

necker Proton Therapy Facility (RPTC) and to validate the production cross-sections of positron emitters based on isotope production yield measurements at this cyclotron-based facility.

The thesis is structured as follows. Chapter 2 provides the physical background of both proton therapy and PET imaging. Moreover, the Rinecker Proton Therapy Center (RPTC), the proton facility whose beamline was modeled in Monte Carlo and at which PET activation experiments were conducted, is presented in section 2.3.

In chapter 3, the Monte Carlo setup used to model the RPTC beamline is shown. The FLUKA Monte Carlo code and the user-routines used in this work are briefly presented in section 3.1 and 3.2. In section 3.3, the conversion formalism from Monitor Units to number of protons is shown and in section 3.4 the influence of the so-called low-dose envelope and the field size on the dose output is analyzed.

The central investigations performed in this thesis, the PET activation experiments, are presented in chapter 4. This chapter starts with the presentation of the materials and methods (section 4.1), followed by quantitative results and studies on the lateral activity profile (section 4.2 and 4.2.3, respectively). An analysis of the performance of the PET scanner concerning its linearity over a wide range of activity concentrations is presented in section 4.3. This chapter is concluded in section 4.4 with the discussion of the uncertainties of the presented activation studies.

Chapter 5 then provides a brief summary of the main results and gives an outlook to possible continuations of this project.

2 Theoretical background of proton therapy and PET imaging

In this chapter, the physical background of radiation therapy with protons (section 2.1), as well as the basics of positron emission tomography (section 2.2) are reviewed. Furthermore, the proton therapy facility, where the activation experiments described in chapter 4 were carried out, is briefly presented in section 2.3.

2.1 Principle of proton therapy

As already mentioned in the introduction (chapter 1), the main advantage of proton therapy as compared to conventional radiotherapy using high-energy x-rays is the favorable depth dose distribution. This is due to the different physical processes governing interaction of radiation with matter. This section therefore gives an insight to the physics related to radiotherapy with heavy charged particles.

2.1.1 Interaction of heavy charged particles with matter

On their passage through matter, heavy charged particle projectiles, like accelerated protons, interact with the traversed tissue via electromagnetic and hadronic interactions, depositing most of their energy at the end of their range in the Bragg peak (BP). The average rate at which a charged particle with energy E loses energy per unit path length $\mathrm{d}x$ is described by the *linear stopping power*

$$S = -\frac{\mathrm{d}E}{\mathrm{d}x} \, . \tag{2.1}$$

In practice, the stopping power is often expressed in units of mass thickness ($\mathrm{MeV\,cm^2/g}$). It is therefore divided by the density of the traversed medium, ρ, resulting in the *mass stopping power* which is not very different for a wide range of materials [Leo, 1994].

In the therapeutically relevant energy range, which is about 50 MeV to 250 MeV for protons, corresponding to a penetration depth of about 2 cm to 38 cm in water, the dominant interaction with matter is via inelastic Coulomb interactions with the target electrons, as it can be seen in figure 2.1. The resulting rate of average energy loss per unit path length

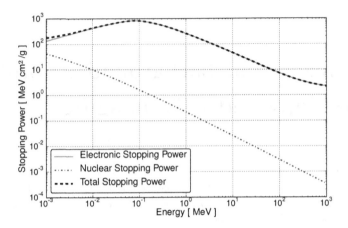

Figure 2.1: Electronic, nuclear and total stopping power of protons in water, data from NIST [Berger et al., 2005]. Energy loss is highly dominated by electromagnetic stopping power, nuclear stopping power is negligible except for small, therapeutically not relevant energies.

for heavy charged particles, also known as electronic stopping power, is described by the Bethe-Bloch formula [Bethe, 1930, Leo, 1994], valid at energies larger than about 1 MeV:

$$-\frac{\mathrm{d}E}{\mathrm{d}x} = 2\pi n_e r_e^2 m_e c^2 \frac{z^2}{\beta^2} \left[\ln\left(\frac{2m_e c^2 \beta^2 W_{\max}}{I^2(1-\beta^2)}\right) - 2\beta^2 - \delta - 2\frac{C}{Z} \right] , \qquad (2.2)$$

where $n_e = N_A \rho \frac{Z}{A}$ is the electron density of the target material, N_A is the Avogadro constant, ρ is the mass density, Z the atomic number and A the atomic mass number of the target material. r_e and m_e are the classical electron radius and the electron rest mass, respectively. z and β are the charge and the velocity of the projectile scaled by the speed of light c, respectively. I is the ionization potential of the medium, whereas $W_{\max}$ is the maximum energy loss in a single collision with a free electron. Two corrections are included in this formulation of the Bethe-Bloch formula - the density effect correction δ which is important at high projectile energies and the shell correction C, important at low energies. Both corrections also depend on the properties of the absorber material.

One can see from eq. (2.2) that the stopping power should increase with decreasing speed β. However, at low energies ($\lesssim 0.1$ MeV), the stopping power drops again, as it can be seen in figure 2.1. In this energy range, recombination processes between the projectile protons and electrons from target atoms occur, which consequently reduce the charge z in eq. (2.2) to an effective charge z_{eff}. The dependence of the effective charge on the particle velocity is described by the semi-empirical formula [Barkas and Evans, 1963]:

$$z_{\mathrm{eff}} = z \left(1 - \exp(-125\beta z^{-\frac{2}{3}})\right) . \qquad (2.3)$$

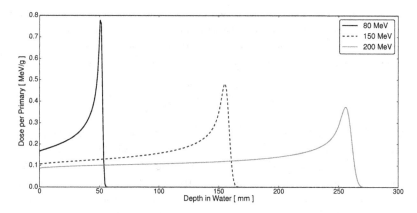

Figure 2.2: Depth-dose profiles of a 80 MeV (black), 150 MeV (dashed) and 200 MeV (gray) proton beam in water (FLUKA simulations). Range straggling increases with larger beam energy and hence penetration depth.

Hence, for the depth-distribution of the ionization density of heavy charged particles, the dominant $1/\beta^2$-dependence in eq. (2.2), combined with the decreasing effective charge of eq. (2.3) at low energies, gives rise to an almost constant plateau, followed by the sharp rise and fall at the end of the ion range, the Bragg-curve.

The most important quantity in radiotherapy is the *absorbed dose*, D. It is a macroscopic quantity, defined as the average energy $\mathrm{d}\epsilon$ imparted by ionizing radiation to matter per unit mass,

$$D = \frac{\mathrm{d}\epsilon}{\mathrm{d}m} \, , \tag{2.4}$$

and is usually expressed in Gray ($1\,\mathrm{Gy} = 1\,\mathrm{J/kg}$). For a mono-energetic radiation field, the absorbed dose can be linked to the stopping power via the primary particle fluence $\Phi = \mathrm{d}N/\mathrm{d}A$, which is the number of particles $\mathrm{d}N$ traversing a sphere of cross-sectional area $\mathrm{d}A$:

$$D = \frac{\Phi}{\rho} \frac{\mathrm{d}E}{\mathrm{d}x} \, . \tag{2.5}$$

This relation only holds when there is secondary electron equilibrium. That means that the amount of energy carried into a small volume of interest by secondary electrons generated outside of the volume is on average equal to the amount of energy transported out of this volume element by secondary electrons created in the volume.

2.1.2 Particle range and range straggling

Another quantity of interest is the range R of a charged particle, which is the expectation value of the path length that the particle follows until it comes to rest [Attix, 1986]. Generally, one uses the continuous slowing down approximation (CSDA) to express this range. In this approximation, continuous energy loss is assumed instead of multiple discrete losses. However, the difference between the actual range R and the CSDA-range R_{CSDA} is negligible [Attix, 1986]. For an initial energy E_0, the range can be expressed in terms of the stopping power:

$$R(E_0) \simeq R_{\mathrm{CSDA}}(E_0) = \int_{E_0}^{0} \left(\frac{\mathrm{d}E}{\mathrm{d}x}\right)^{-1} \mathrm{d}E \ . \tag{2.6}$$

Due to multiple Coulomb scattering (cf section 2.1.3), the particles follow a zigzag path through the absorber. A different definition of range is the so-called projected range R_{p}, which is the average penetration depth in medium along the particle's initial direction. The projected range, defined as straight line, is therefore always smaller than the total zigzag path length. However, the effect of lateral scattering is small for heavy charged particles, so that R_{p} and R_{CSDA} do not differ significantly [Leo, 1994].

Yet, the number of collisions experienced by each individual beam particle, as well as the energy transferred at each collision are subject to statistical fluctuations. The small range spread around the mean range caused by these fluctuations is called *range straggling* or *energy straggling*. It is the reason for the increased width of the Bragg peak as compared to calculations using the average energy loss of a single particle. For protons in tissue, the difference is in the order of 1 % of the mean range [Linz, 2011]. Range straggling therefore becomes more significant for larger beam energies, which can be seen in figure 2.2.

The energy loss and range distribution can be assumed to be Gaussian for thick absorbers, with a width proportional to $1/\sqrt{M}$, where M is the mass of the projectile [Schardt et al., 2010]. The BP width for heavier ions is therefore smaller than for protons.

In practice, the BP is also broadened by the unavoidable momentum dispersion $\Delta p/p$ of the beam before entering the medium, due to the accelerator and beamline elements. For clinically used accelerators and beamlines, this initial momentum spread is comparable to the effect of range straggling in tissue, and is taken into account in the Monte Carlo setup of the clinical beam model described in chapter 3.

2.1.3   Lateral scattering

In addition to the longitudinal spread discussed so far, the ion beams also broaden in lateral direction. This is especially pronounced for lighter ions, like protons. The main reason for lateral scattering is caused by elastic Coulomb interactions with the target nuclei [Schardt et al., 2010]. In contrast, deflections of the primary beam due to interactions with the target electrons are negligible due to the much smaller electron mass.

An analytical solution for multiple Coulomb scattering (MCS) was derived by Moliere [1948]. Ignoring the small probability of large-angle single scattering ($> 10°$), the angular distribution of a large number of independent scattering events can be well approximated

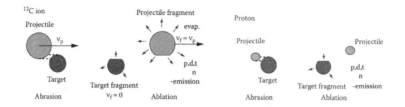

Figure 2.3: Sketch of the abrasion-ablation model for nuclear collisions of ^{12}C (left) and proton (right) projectiles with target nuclei. From Parodi [2011a], adapted from Schardt et al. [2010].

by a Gaussian distribution with a standard deviation [Highland, 1975]:

$$\sigma_\theta[\text{rad}] = \frac{14.1\,\text{MeV}}{\beta pc} z \sqrt{\frac{d}{L_{\text{rad}}}} \left[1 + \frac{1}{9}\log_{10}\left(\frac{d}{L_{rad}}\right)\right] , \qquad (2.7)$$

where β and p are the incident particle velocity and momentum, respectively, and the absorber material specific thickness d and radiation length L_{rad}. From the $(\beta pc)^{-1}$ term in eq. (2.7) it follows, that multiple Coulomb scattering increases with decreasing particle energy.

2.1.4 Nuclear reactions and positron emitter production

Heavy charged particles can also interact with the target nuclei via the strong nuclear force. These nuclear interactions do not have to be confused with the nuclear stopping power briefly mentioned in section 2.1.1, which is due to interactions with the electromagnetic field of the target nuclei.

These nuclear collisions may result in fragmentation of target nuclei and, if heavier than protons, also of projectile nuclei. Peripheral collisions are the most frequent reactions due to geometrical reasons and can be described as a two-step process by the so-called *abrasion-ablation model*. Figure 2.3 illustrates the fragmentation reactions for ^{12}C ions (left) and protons (right). In a first step, the collision, partially excited prefragments are produced, which then de-excite in a second step by nucleon evaporation or photon emission (ablation).

Projectile fragments continue to penetrate with approximately the same velocity and direction as before. While projectile fragments with the same charge number but a lower mass number than the primary beam particles have shorter ranges than the initial particles, fragments with lower charge number have longer ranges and deposit energy far behind the Bragg peak of the initial particle species. In contrast, target fragments almost stay at rest because of their heavy mass and reaction kinematics.

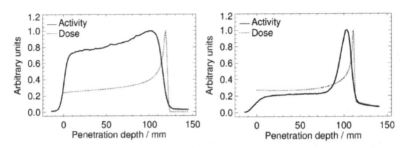

Figure 2.4: Example of depth activity distributions for proton (left) and carbon (right) irradiation of a PMMA sample. From Parodi [2004].

Because of the nuclear interactions, the initial particle fluence Φ_0 reduces with increasing depth x:

$$\Phi(x) = \Phi_0 \, e^{-N\sigma_R x} \ . \tag{2.8}$$

N is the nuclear density of the medium and σ_R is the total reaction cross section for nucleus-nucleus collisions. According to eq. (2.5), this consequently reduces the dose delivered by the primary ions with increasing depth, while secondary particles from these nuclear reactions additionally contribute to the delivered dose. A negative consequence of nuclear interactions is an increased lateral spread of narrow ion beams (cf section 3.4).

Part of the fragments produced in nuclear interactions may be β^+-radioactive isotopes. In the case of ^{12}C-ion beams, positron-emitting projectile fragments like ^{11}C and ^{10}C may be generated and penetrate the target until they stop shortly before the primary stable beam particles, according to the ratio between their atomic mass number and the proton number, A/Z^2. The activity distribution due to these radioactive projectiles, namely the β^+-activity maximum shortly before the Bragg peak (cf figure 2.4, right), can be measured by means of positron emission tomography [Parodi, 2004].

In the proton case, only target fragments may contribute to the β^+-activation. They are produced along the beam path until the projectile's energy drops below the threshold of nuclear reactions, which is around 10 MeV to 20 MeV, depending on the target nucleus. This corresponds to a maximum of the activity distribution at a few millimeters before the Bragg peak (cf figure 2.4, left).

Thus, the by-products of irradiation with heavy charged particles, namely the β^+-radioactive fragments of projectile and target nuclei, have the potential to be used for non-invasive treatment verification, without increasing the dose to the patient's healthy tissue. The detection method used for monitoring the irradiation induced β^+-activity is positron emission tomography and will be presented in the following section.

2.2 Principles of PET imaging

The original purpose of PET (positron emission tomography) in medicine is an imaging technique used in nuclear medicine. It is a non-invasive method to image physiological processes in living tissue. So-called radioactive tracers (e.g. ^{18}F-FDG), are injected to the patient and enrich in malignant tissue. The resulting activity distribution can then be detected and allows volumetric imaging of the physiological function.

However, PET-scanners can in principle be used to measure any kind of β^+-activity distribution, like irradiation-induced activity (cf section 2.1.4). They are therefore a promising tool for dose monitoring in ion beam therapy. This section aims to give an insight to the physical and technical background of PET imaging.

2.2.1 Physical background of PET imaging

The decay process making PET imaging possible is the β^+-decay, or positron decay. In unstable, neutron deficient nuclei, a proton is converted to a neutron under the emission of a positron e^+ and an electron neutrino ν_e in the following reaction:

$$_Z^A X \ \rightarrow \ _{Z-1}^A Y + e^+ + \nu_e \ . \tag{2.9}$$

Since the proton mass is lower than the neutron mass, this process cannot occur in isolated protons. Due to conservation of energy, the binding energy of the daughter nucleus must be higher than the binding energy of the mother isotope. Thus, the amount of energy released by that reaction is

$$Q = \left(M(_Z^A X) - M(_{Z-1}^A Y) - 2m_e \right) c^2 \ , \tag{2.10}$$

where $M(_Z^A X)$ and $M(_{Z-1}^A Y)$ are the atomic masses of the mother and the daughter isotopes, respectively, and m_e is the electron mass[1]. This energy goes into kinetic energy of both positron and neutrino, resulting in a continuous energy spectrum of the positron to the maximum possible energy values [Magill and Galy, 2004].

For the positron emitters studied in this work, ^{11}C, ^{13}N and ^{15}O, the positron's maximum possible energies are 0.97 MeV, 1.20 MeV and 1.74 MeV, respectively. This corresponds to a maximum positron range in water of 3.8 mm, 5.0 mm and 8.0 mm, respectively. The average β^+- ranges in water are between 0.85 mm and 1.80 mm for these positron emitters [Saha, G.B., 2010].

While the neutrino does not interact and escapes the system, the β^+-particle loses its energy in multiple Coulomb interactions with orbital electrons of the absorber atoms, suffering several angular deflections. When it is almost at rest, it forms an unstable bound state with an electron of the medium, the so-called positronium. Depending on the relative

---

[1]If the daughter nucleus is produced in an excited state, also the excitation energy E^* has to be subtracted from the right hand side of the eq. (2.10).

True coincidence   Random coincidence   Scattered coincidence

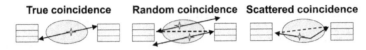

Figure 2.5: True (left), random (middle) and scattered (right) coincidence events detected by two detectors. The location of the annihilation is marked by the star, the solid lines show the real pathways of the photons whereas the dashed lines indicate the lines-of-response (LORs) which is assigned by the detectors. From Parodi [2004].

spin states of the two particles, the singlet state para-positronium or the triplet state ortho-positronium are created. Para-positronium annihilates into two photons within 125 ps, whereas ortho-positronium decays within 140 ns preferably into three gammas. Due to the large difference in lifetime, the 3γ-emission can be neglected in practice [Parodi, 2004]. Thus, only the 2γ-emission is considered.

The two annihilation photons have an energy of 511 keV, corresponding to the electron and positron rest mass, and are ejected in almost opposite direction according to momentum conservation. However, the small residual momentum of the positron-electron system before the annihilation process results in a small deviation from the exact 180° of at most ±0.25° [Saha, G.B., 2010].

The fundamental of PET is the detection of the 511 keV pairs, emitted isotropically in 4π, by two opposite detectors within a given coincidence time window (typically a few nanoseconds). The straight lines connecting the centers of the two detectors are called lines-of-response (LORs) that can be understood as the lines, along which the annihilations occurred. All coincidence events within this time window are called prompts and can be classified as *true*, *random* and *scattered* events [Saha, G.B., 2010]. The three types of coincidence events are illustrated in figure 2.5.

The favored coincidence events are of course true events. They arise, when two gammas from the same annihilation event are detected along the LOR (figure 2.5, left). The second type of events are random events. They are coincidences that occur, when two gammas from two independent annihilation events are detected within the same time window (figure 2.5, middle). If one or both photons from one annihilation event suffer angular deflection due to Compton scattering in the object and are detected within the same coincidence time window, one speaks of scattered events (figure 2.5, right). The energy loss of the photons may not be to large, because otherwise it will not fall within the energy window of the detector electronics.

Since the PET scanner cannot distinguish between the three types of coincidence event, the latter two events raise the background of the reconstructed image, degrading the image contrast [Saha, G.B., 2010]. There is still another physical process that worsens the image quality. Photon attenuation due to absorption and scattering within the scanned object results in a loss of detected events. There are several methods aiming to correct these artifacts. In section 4.1.2.2, the correction methods used for image reconstruction in this work are presented.

2.2.2 PET scanning systems

Typical PET scanners are composed as follows. Several small inorganic scintillation detectors are created by partially cutting a large block of detector crystal, forming so-called block detectors. These block detectors are typically about 3 cm thick and coupled to photomultiplier tubes. These blocks are, together with the necessary electronics, arranged in full-rings arrays, with diameter of 80 to 90 cm [Saha, G.B., 2010].

The choice of the scintillation detector material has a large influence on the quality of the resulting image. First of all, a high stopping power of the detector material is desired. Since the probability of the photoelectric effect is proportional to $\propto Z^5/E_\gamma^3$, high Z and high density materials are favorable. Moreover, the scintillation decay time of the crystal should be short. The longer a scintillator crystal needs from the excitation by the 511 keV photon until it decays to its ground state, emitting visible light, the larger the dead time of the detector gets. Fast scintillation decay time is also important to reduce the coincidence time window and thus, reduce random background. Further important characteristics are a high photon yield (per keV) and the energy resolution of the detector. A low light output results in poor energy resolution [Saha, G.B., 2010].

The commonly used detector materials are BGO (bismuth germinate), GSO (gadolinium oxyorthosilicate), LSO (lutetium oxyorthosilicate) and LYSO (lutetium yttrium oxyorthosilicate). The latter two detectors combine a high photon attenuation and a high light output, together with a short scintillation decay time (40 ns). However, this detector has the disadvantage to contain the radioactive isotope ^{176}Lu. This is typically not a problem for imaging applications in nuclear medicine, because its photon energy ranges only up to 400 keV [Saha, G.B., 2010].

Due to the lack of anatomical information in functional imaging, it is beneficial to combine PET scanners with computed tomography X-ray scanners (CT). In diagnostics and radiation therapy planning, these PET/CT scanners have the advantage to provide both anatomical and functional imaging. Another advantage is the possibility to use the acquired CT image as a basis for the calculation of the attenuation and scattering correction of the annihilation photons from the PET image.

2.3 The Rinecker Proton Therapy Center (RPTC)

The Rinecker Proton Therapy Center, RPTC, is a clinical proton therapy facility in Munich which was opened for patient treatment in March 2009. A detailed Monte Carlo model of the facility's beamline is described in chapter 3 and in Englbrecht [2014]. The activation experiments described in chapter 4 were performed in this treatment site as well. This section shall therefore give a brief overview on the facility, highlighting the features relevant for the presented studies.

The proton beam is created by a superconducting cyclotron (figure 2.6 (a)), where the particles are accelerated up to 250 MeV. Besides the much smaller size, a major advantage

(a) Cyclotron   (b) Energy selection system

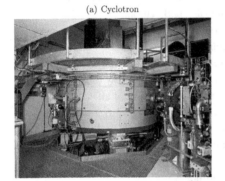   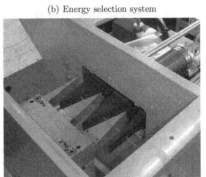

(c) Proton gantry

Figure 2.6: Cyclotron, energy selection system and gantry (from the backside) of the RPTC. Reprinted by permission from RPTC [rpt].

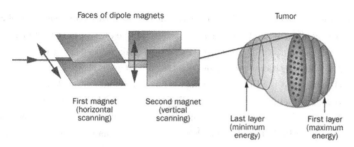

Figure 2.7: Principle of pencil beam scanning. Two perpendicular magnets deflect the pencil-like proton beam in horizontal and vertical direction and allow to scan across a thin slice of constant depth. Different depth layers can be covered by changing the beam energy. Reprinted by permission from Macmillan Publishers Ltd: Nature Reviews Clinical Oncology [Durante and Loeffler, 2010], © 2010.

of this cyclotron as compared to synchrotrons is the quasi-continuous beam extraction with steady intensity [Krischel, 2011]. This reduces irradiation time tremendously and therefore has a positive impact on the PET activation experiments, as well as their data evaluation.

After leaving the cyclotron, the proton beam energy is degraded by graphite wedges in the beam path (figure 2.6 (b)) to the desired treatment energy (75 to 245 MeV) and then guided to one of the treatment rooms. The beam is transported in vacuum, and magnetic lenses keep the beam focused until it reaches the gantry. The four proton gantries (figure 2.6 (c)) can be rotated around their horizontal axis and provide 360° of beam coverage around the patient [Jongen, 2011].

Dose is delivered via active spot-scanning (cf figure 2.7). Narrow, pencil-like proton beams are deflected magnetically in both horizontal and vertical direction, covering thin slices of the treatment volume. By successively reducing the beam energy from the maximum down to the lowest desired value, i.e. from the most distal to the most proximal slice, respectively, target volumes of arbitrary shape can be irradiated. This demanding technique allows position dependent dose variation resulting in better dose conformity in the target volume while reducing the dose to the surrounding tissue. Moreover, there is less material in the beam path as compared to passive beam shaping, where patient-specific compensators and collimators are used. Consequently, less unwanted secondaries (mainly neutrons) are generated upstream to the to patient [Schardt et al., 2010]. In section 3.2, the Monte Carlo implementation of this pencil beam scanning system is described.

A commercial full-ring PET/CT scanner (Philips Gemini 16 Power[2]) is installed in a separate nearby room at about 2 min walking distance. 17 864 GSO crystals, each $4 \times 6 \times 20\,\text{mm}^3$ small, are coupled to 420 photo-multiplier tubes. The detector ring diameter is 90 cm and the axial field-of-view (FOV) is 18 cm. The spatial resolution in the center is 5.0 mm and 4.9 mm in axial and transverse direction, respectively. Since it is a combined PET/CT scanner, it allows CT based attenuation and scattering correction of the annihilation photons. Furthermore, for treatment verification, it offers the possibility

---

[2]Koniklijke Philips N.V., Eindhoven, The Netherlands

to compare the combined PET/CT image with the CT image used for treatment planning. Both treatment room and PET/CT scanner are equipped with a laser alignment system indicating the location of the treatment isocenter. This is used for accurate phantom positioning in the activation experiments described in chapter 4.

Due to the fact that the PET/CT scanner is not inside the treatment room, only offline PET/CT treatment verification is possible. This has the drawback, that many short-lived isotopes cannot be detected anymore, resulting in an overall worse signal as compared to in-room, or in-beam PET [Parodi et al., 2008]. On the other hand, unlike custom-made in-beam PET solutions, commercial PET/CT scanners are available at relatively low cost and can be used for diagnostic purposes as well. Moreover, the offline PET concept does not prolong the time the treatment room is occupied, resulting in a higher patient throughput for irradiation.

3 Monte Carlo modeling of the clinical proton beam

A detailed description of the Monte Carlo modeling of the clinical proton beam which is used for the investigations in chapter 4 can be found in Englbrecht [2014]. However, there are still two important aspects missing which need to be implemented before the Monte Carlo beamline model can be used for simulations that can be compared to predictions by the treatment planning system and measured data. Therefore, after a brief description of the FLUKA Monte Carlo code, section 3.2 gives an insight on the implementation of an improved geometrical characterization of the spot scanning system in the user-routine source.f. The dose calibration of the Monte Carlo output is presented in section 3.3. Finally, the influence of the so-called low-dose envelope of scanned proton beams is studied in section 3.4.

3.1 The FLUKA Monte Carlo Code

FLUKA (FLUktuierende KAskade) is a multipurpose Monte Carlo (MC) code for the calculation of particle transport and interaction with matter [Battistoni et al., 2007, Ferrari et al., 2005]. While the original code was written in the sixties and its purpose was to design shielding in high energy proton accelerators, FLUKA was gradually further developed and covers nowadays a large variety of applications, including neutrino physics, cosmic ray physics, calorimetry, radiation protection, dosimetry, hadron therapy, activation studies and many more.

The current version, 2011.2b.6, supports interactions and transport of more than 60 different elementary particles and all kinds of heavy ions in a large energy range. Photons and electrons can be simulated from 1 keV up to thousands of TeV, hadrons are supported up to 20 TeV and neutrons down to thermal energies [Battistoni et al., 2007].

Arbitrarily complex geometries can be handled using combinatorial geometry. One feature that makes FLUKA particularly interesting for radiotherapy is its possibility to handle voxel geometries[1]. CT images can be imported and converted to a voxel file where

---

[1]Any geometry can be described in terms of equally sized, small parallelepipeds *(voxels)* forming a three-dimensional grid [FLU, 2013]. Voxel is a portmanteau word of the words *volume* and *pixel*.

physical properties are attributed to each voxel, based on the corresponding Hounsfield Unit (HU) of the CT image.

3.1.1   The FLUKA input file

To start a FLUKA simulation, an ASCII input file, containing several *commands*, each consisting of one or more *lines* (= *cards*) has to be created. Each card contains one keyword, followed by six floating point values (*WHATs*) and one character string (*SDUM*). In this input file, the user has to specify the geometry of the setup (GEOBEGIN, SPH, XYP, VOXELS, REGION, ..., GEOEND cards), can define materials, their composition and properties with the MATERIAL, COMPOUND, MAT-PROP and STERNHEI cards and assign them to certain regions using the ASSIGNMA card. The particle type and energy, as well as the source position has to be defined in the BEAM and BEAMPOS cards. The user can add several *detectors* to the input files (e.g. USRBIN card) and specify, which physical quantity should be scored and where. Moreover, some physics settings and production or transport thresholds can be set and modified (PHYSICS, DELTARAY, ... cards). Different types of biasing can also be activated but has not been necessary throughout this thesis.

The graphical user interface FLAIR [Vlachoudis, 2009], based on python and Tkinter, facilitates editing error-free FLUKA input files by displaying the input file in an easily readable way and allows to compile and start FLUKA jobs from a GUI environment.

3.1.2   FLUKA user-routines

For the purpose of this work, modifications in the input file were not sufficient. Therefore, a number of user interface routines, all written in Fortran 77, can be modified by the user in order to alter simulation input and output, and in limited cases even particle transport. For activating these user-routines, they have to be compiled and linked to the FLUKA library and have to be called by a special command in the FLUKA input file.

The following user-routines were used in this work:

source.f

This user-routine, activated by the SOURCE card, allows to sample primary particle properties from distributions in e.g. space and energy, which are too complex to be described by the FLUKA cards [Battistoni et al., 2007]. This is essential for the setup of the clinical proton beamline. In section 3.2, our adaptation of source.f is explained in detail.

fluscw.f

The user-routine fluscw.f allows the user to modify the simulation output and is activated by the USERWEIG card. Any fluence, calculated with USRBIN can be altered during runtime with user-specified functions. This makes it possible to weight the output according to particle type, energy, generation etc. For the studies presented in chapter 4,

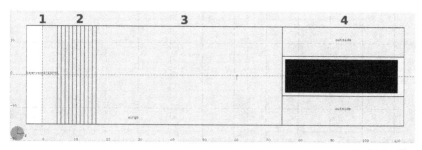

Figure 3.1: FLUKA model of the RPTC beamline, as described in Englbrecht [2014]. The proton beam, coming from the left, exits the vacuum (white) through the vacuum window (1), crosses the multi strip ionization chamber (MSIC) and transmission ionization chamber (TIC) (2) and propagates through the air-gap (3) between the nozzle and the patient table, where arbitrary phantoms can be placed (4).

this user-routine is used to compute the amount of produced positron emitters by combining the proton fluence with energy-dependent cross-sections for the desired proton-induced nuclear reactions.

usrrnc.f, usrini.f, usrout.f

For calculating the positron emitter yield using FLUKA's internal model, these three subroutines are necessary. If the USERWEIG card is used in the input file, the routine usrrnc.f is called every time a residual nucleus is produced. The subroutines usrini.f and usrout.f are called every time the USRICALL and USROCALL card are found in the input stream, respectively. They are used to initialize estimators for the scoring of requested positron emitter yields and write the corresponding data into these USRBINS.

3.1.3 FLUKA model of the beamline

The scope of this subsection is to recall the geometrical modeling of the clinical beamline, which can be found in [Englbrecht, 2014, 4.1]. Figure 3.1 illustrates the components which are permanently installed and explicitly included in the simulation, since these elements, as well as the air-gap between nozzle and the patient table, have an influence on the spot size and the proton range. The phantom in this figure represents one of the phantoms used in the activation studies in chapter 4 and can be replaced by arbitrary other phantoms or CT voxel geometries.

3.2 The spot scanning routine

A realistic clinical proton beam source cannot be described by a simple point source, as it would be the default source in a FLUKA simulation. Hence, an extended proton source has to be sampled, where the initial spot is described by a two-dimensional Gaussian with two independent FWHMs. This resulting ellipsis can be tilted by a user-specified angle. The

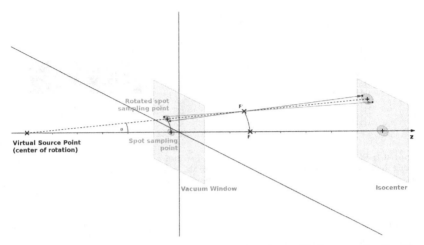

Figure 3.2: Sketch of the geometry of the spot scanning as it is implemented in the `source.f` routine. The spot sampling point, which is located shortly before the origin of the coordinate system and the vacuum window, as well as the focal point F are rotated by an angle α. The rotated spot sampling point and focal point F' thus define the direction cosine of the created particles.

corresponding values are based on measurements performed at the RPTC and are stored in the data file `spots`. Moreover, the clinical proton source is not purely mono-energetic, thus the introduction of a momentum spread $\Delta p/p$ of the initial proton beam momentum is required. The data for the momentum spread is stored in the file `pspread`. Also, the kinetic energy of the initial proton beam has to be raised by a constant offset, corresponding to a water equivalent path length of 0.162 mm in order to match experimentally measured beam ranges in water. This requires a table (`dEdx`) where the stopping powers in water are saved for the complete clinical energy range.

For this purpose, the user routine `source.f` was programmed [Englbrecht, 2014, 3.2.4 and 4.2.2]. However, the source-routine did not include the possibility to deflect the initial proton beam in a way to simulate spot scanning (cf section 2.3). It therefore had to be extended by this feature. This section shall give an insight to the implemented geometrical considerations.

By the time this user-routine was written, the effects of the focusing magnets on the resulting proton beam in the vicinity of the isocenter have not fully been investigated yet. As it turned out, beam focusing (in analogy to focusing in ray optics), can be neglected in this present clinical model [Englbrecht, 2014, 4.4.3]. Nevertheless, the current version of `source.f` still offers the possibility to account for focusing of the proton beam, and the focal length can be read from the input table `spots`. In order to make beam focusing negligible, as it is done throughout this work, the focusing parameter f is set to 10^8 cm. A sketch of the relevant planes and coordinates can be seen in figure 3.2.

Since the origin of the coordinate system used in the FLUKA model is not the center of rotation but the location of the vacuum window, the coordinates of the sampling point, the focal point and the isocenter have to be shifted in z-direction by a shift $\delta = z_{air} - z_{SID}$. Here, z_{air} is the total length of traversed air between the vacuum window and the treatment isocenter and z_{SID} is the virtual source-to-isocenter distance, which is specified by the facility.

After this translation, a rotation around a line through the origin of the coordinate system is performed. Its direction and orientation is given by the normalized outer vector product of the position of the isocenter relative to the virtual source position[2] $(0, 0, z)^T$ and the target position of the spot in the isocenter plane $(x, y, z)^T$, resulting in the unit vector

$$
\vec{u} = \frac{1}{z \cdot \sqrt{x^2 + y^2}} \begin{pmatrix} -yz \\ xz \\ 0 \end{pmatrix} .
\tag{3.1}
$$

In case there is no deflection of the beam, this vector is replaced by a zero-vector in order to prevent division by zero in the following calculations.

With this direction, the rotation matrix becomes

$$
R_{\vec{u}}(\alpha) = \begin{pmatrix} 2u_1^2 \sin\frac{\alpha}{2} + \cos\alpha & 2u_1u_2 \sin\frac{\alpha}{2} - u_3 \sin\alpha & 2u_1u_3 \sin\frac{\alpha}{2} + u_2 \sin\alpha \\ 2u_2u_1 \sin\frac{\alpha}{2} + u_3 \sin\alpha & 2u_2^2 \sin\frac{\alpha}{2} + \cos\alpha & 2u_2u_3 \sin\frac{\alpha}{2} - u_1 \sin\alpha \\ 2u_3u_1 \sin\frac{\alpha}{2} - u_2 \sin\alpha & 2u_3u_2 \sin\frac{\alpha}{2} + u_1 \sin\alpha & 2u_3^2 \sin\frac{\alpha}{2} + \cos\alpha \end{pmatrix} ,
\tag{3.2}
$$

where the cosine and the sine of the total deflection α with respect to the origin are

$$
\cos\alpha = \frac{z}{\sqrt{x^2 + y^2 + z^2}} \quad \text{and} \quad \sin\alpha = \frac{\sqrt{x^2 + y^2}}{\sqrt{x^2 + y^2 + z^2}} .
\tag{3.3}
$$

Now, the rotation matrix is applied on the source point $\vec{r_0}$ of the particle $\vec{r_0}' = R\vec{r_0}$, as well as on the focal point $\vec{f}' = R\vec{f}$. The direction cosines $\vec{v}$ of the particle[3] is then the normalized difference of the focal point and the source point:

$$
\vec{v} = \frac{1}{\sqrt{|\vec{f}'|^2 + |\vec{r_0}'|^2}} (\vec{f}' - \vec{r_0}') .
\tag{3.4}
$$

In a last step, the z-coordinate of the new source point is translated back, using the shift $-\delta$.

---

[2]This virtual source-to-isocenter distance is the center of rotation of the deflected proton beam and is given by the vendor. It must not be confused with the sampling point of the extended proton source of the simulation, which is located 3 cm upstream of the vacuum window.

[3]In the FLUKA user routines, the variable names of the source coordinates of the I-th stack particle are XFLK(I), YFLK(I) and ZFLK(I), whereas the direction cosines are called TXFLK(I), TYFLK(I) and TZFLK(I).

3.3   Monitor Unit calibration

The output of a treatment machine is usually measured in so-called *Monitor Units* (MU). For this purpose, monitor chambers (vented ionization chambers) are mounted in the treatment head to measure the ionization induced by the proton beam and thus control the amount of dose delivered to a phantom or a patient. The output dose per MU of the clinical beam is well calibrated under certain conditions and depends on the beam energy. However, this is only a relative quantity and the underlying physical quantity is the proton fluence. This is the quantity of interest for this study, because FLUKA MC simulations can calculate dose distributions and positron emitter yields only *per primary beam particle*.

Thus, accurate knowledge of the conversion between MU and number of delivered protons is essential for quantitative predictions from MC simulations. Unfortunately, the number of protons cannot be calculated directly from monitor units, which made calibration experiments necessary. The setup for these experiments is described in detail in [Englbrecht, 2014, 5.2.2]. In the following subsection, a brief summary of the experiment and the conversion formalism is given before in section 3.3.2 an improved calibration approach is presented.

3.3.1   MU calibration experiments, conversion formalism and validation

A Bragg peak (BP) chamber (type 34070, PTW-Freiburg, Germany) was placed prior to a water tank and was irradiated by a single pencil-like proton beam at different energies, covering the whole treatment energy range. For each energy, 100 MU were irradiated at low dose rate in order to exclude saturation effects of the BP chamber. The measured charge $Q(E)$ at beam energy E can then be converted to the proton number per 100 MU, $N_p^{100\text{MU}}(E)$:

$$N_p^{100\text{MU}}(E) = \frac{Q(E) \cdot W_e}{\left(\frac{\mathrm{d}E}{\mathrm{d}x}\right)|_{air,E_{4mm}} \cdot d} . \tag{3.5}$$

W_e is the average energy that is required to produce an electron-ion pair in air by proton irradiation. Following the recommendations of the ICRU 59 report [ICRU, 1998], the value of humid air, $W_e = 34.8\,\text{eV}$, has been used in Englbrecht [2014]. The geometrical thickness of the chamber's sensitive volume is $d = 2\,\text{mm}$. $\left(\frac{\mathrm{d}E}{\mathrm{d}x}\right)|_{air,E_{4mm}}$ is the stopping power of the air in the BP chamber for a beam energy E. Since the protons have to traverse the entrance window of the ionization chamber (IC) with a water equivalent path length (WEPL) of 4 mm prior to the sensitive measurement volume, the beam energy E is degraded to the energy E_{4mm}. The corresponding new stopping powers in water, $\left(\frac{\mathrm{d}E}{\mathrm{d}x}\right)|_{water,E_{4mm}}$ have been extracted from FLUKA MC simulations. To this aim, the dose at 4 mm in water was scored using a cylindrical USRBIN scoring volume with the radius $r = 4.08\,\text{cm}$, which is the radius of the sensitive volume of the BP chamber. With the density of water at room temperature in the treatment room, $\rho_{water} = 0.99777 \, \mathrm{g/cm^3}$, the stopping power becomes

$$\left(\frac{\mathrm{d}E}{\mathrm{d}x}\right)|_{water,E_{4mm}} = D|_{4mm} \cdot r^2 \pi \rho_{water} \, . \tag{3.6}$$

Now, either the thickness of the chamber's sensitive volume has to be converted into its water equivalent thickness, or the stopping-power in water has to be converted to stopping power in air. Here, we used the latter approach:

$$\left(\frac{\mathrm{d}E}{\mathrm{d}x}\right)|_{air,4mm} = \left(\frac{\mathrm{d}E}{\mathrm{d}x}\right)|_{water,4mm} \cdot \frac{\rho_{air}}{\rho_{water}} \cdot \frac{1}{S_{water,air}} \, , \tag{3.7}$$

where $\rho_{air} = 1.20484 \, \mathrm{kg/m^3}$ is the density of air in the vented IC and $S_{water,air} = 1.137$ is the mass stopping power ratio from water to air for protons [IAEA, 2000], which in first approximation does not depend on the energy.

Of course, the MU to proton number conversion has to be validated before usage. For this purpose, a homogeneous spread out Bragg peak (SOBP) in a cubic PMMA phantom was planned and optimized using the treatment planning software XiO® (Elekta AB, Stockholm, Sweden). The planned dose cube consisted of 10 different energy layers so that the SOBP was ranging from 15 cm to 20 cm depth and had a lateral field size of $6 \times 6 \, \mathrm{cm^2}$. In the peak region, the comparison between the planned dose and the simulated dose yielded a mean difference of 2.7 %, which is not satisfactory for certain applications, like re-calculating the absolute dose of a treatment plan or the PET activation experiments described in the chapter 4. Therefore, in this work the existing calibration approach was further improved and validated, as described in the following subsection.

3.3.2 Further improvement of the MU calibration

The main sources of uncertainties had been pointed out to be the W_e-value and the fluence correction factors that had to be taken into account, because the dose cube is calculated in PMMA [Englbrecht, 2014]. The latter source of uncertainty can be easily eliminated by using water instead of PMMA. A dose cube with the same lateral and distal dimensions as described above has therefore been planned in a water phantom. The dose in the SOBP was planned to be 6 Gy(RBE), corresponding to 5.455 Gy[4]. The plan consisted of 12 energy layers from 143.75 MeV to 174.33 MeV and the spot spacing between each spot was chosen to be 3 mm, totaling 2864 spots for the complete plan.

The plan was recalculated in FLUKA using different MU to proton number conversion strategies. In a first approach, the same method as described for PMMA was used. A second recalculation was performed with $W_e = 34.23 \, \mathrm{eV}$ for dry air near sea level, as recommended by the TRS report 398 [IAEA, 2000]. Next, the mass stopping power

---

[4]In order to take into account the different biological effectiveness of proton irradiation as compared to x-ray irradiation, the physical dose (Gy) is multiplied by the relative biological effectiveness of 1.1, resulting in the equivalent dose GyE or Gy(RBE).

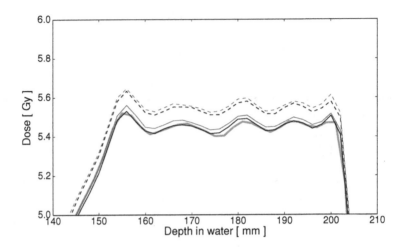

Figure 3.3: The SOBP was planned using the planning software XiO (thick curve). The recalculated plan using the mass stopping power ratio $S_{water,air} = 1.137$ is described by the thin gray curves, while the black lines depict the recalculation using the energy dependent mass stopping power ratio, based on the FLUKA output of the corresponding stopping powers. The W_e-value in the conversion is set to 34.23 eV (solid lines) and to 34.8 eV (dashed lines). The doses are laterally integrated over $2 \times 2\,cm^2$ along the central axis.

ratio $S_{water,air} = 1.137$ was questioned. This constant value was replaced by an energy-dependent mass stopping power ratio

$$S_{water,air}^{\text{FLUKA}}(E) = \frac{\rho_{air}}{\rho_{water}} \cdot \frac{\frac{dE}{dx}|_{water}(E)}{\frac{dE}{dx}|_{air}(E)}, \tag{3.8}$$

with the same two densities ρ_{water} and ρ_{air} as described above. By activating the option PRINT in the FLUKA's DELTARAY-card, the stopping power tables precalculated and used by FLUKA are printed out to the simulation output file. Based on these stopping powers, the mass stopping power ratio water-to-air, $S_{water,air}^{\text{FLUKA}}(E)$, takes values from 1.1322 up to 1.1348 in the clinical energy range.

In the peak region, excellent agreement between predicted and recalculated dose height could be achieved using the energy dependent mass stopping power from eq. (3.8) and the W_e-value for dry air, $W_e = 34.23$ eV (figure 3.3, solid black line). The mean dose in the peak region differs by only 0.03 %. Using $W_e = 34.8$ eV instead overestimates the mean dose in the peak region by 1.7 %. The difference is even larger when using the constant mass stopping power ratio (figure 3.3, gray). Based on this results, the W_e-value for dry air and the energy-dependent mass stopping power was used in the following.

However, when the plateau region of the depth dose profile is regarded, the perfect

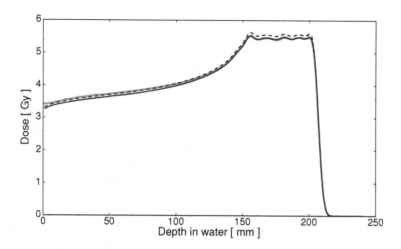

Figure 3.4: Complete depth-dose profile of the dose cube. The line scheme is the same as used in figure 3.3.

agreement disappears. The dose predicted by XiO in the plateau region is larger than the dose calculated in FLUKA, no matter which MU conversion was used (figure 3.4). Although the maximal difference along the plateau region is still rather small ($< 3\%$ for all cases), it is yet larger than the desired precision. The perfect agreement in the SOBP region and the difference in the plateau region indicate, that there is a different effect playing a role than the MU calibration itself. An explanation for this discrepancy can be found when the lateral shape of individual pencil beams from the MC simulation and from the TPS is analyzed in more detail. The following section therefore deals with the so-called *low-dose envelope* of scanned proton beams.

Note, that for the evaluation the dose was laterally integrated over $2 \times 2\,cm^2$ along the dose cube's central axis. This has been done because on the one hand, comparing single voxels or slices would not be robust against small binning effects. On the other hand, integrating over a too large area would further increase the influence of the low-dose envelope.

3.4 Low-dose envelope and Field Size Factor

Even several centimeters from a scanned proton pencil beam's central axis, low doses can be detected. This effect has been first described as *beam halo* [Pedroni et al., 2005] and later as *low-dose envelope* [Sawakuchi et al., 2010a,b]. For scanned proton beams with energies larger than about 150 MeV, mainly secondary particles created in non-elastic nuclear interactions in the phantom are responsible for the low-dose envelope. At lower energies, the probability for non-elastic nuclear interactions in the phantom is small. In this case, the low-dose envelope is dominated by primary protons which have undergone large-angle elastic scattering in the phantom and upstream beamline components [Sawakuchi et al., 2010a].

Although the low-dose envelope of an individual pencil beam is orders of magnitude smaller than the dose in the beam's central axis, up to 15 % of the total dose delivered in a complete plan can be attributed to the low-dose envelope [Pedroni et al., 2005]. This is because of the large number of individual pencil beams needed to irradiate the target volume and the large extent of the pencil beam's low-dose envelope.

In order to evaluate whether the influence of the low-dose envelope can explain the observed discrepancies in the validation of the MU calibration (section 3.3.2), we compared the lateral profiles of individual pencil beams calculated with the treatment planning software XiO to the profiles of the simulated pencil beams. This is followed by a study of the field size dependence of the output factor. In order to investigate the influence on clinically more relevant cases, section 3.4.3 presents a comparison of TPS calculation, MC simulation and measured dose for different sizes of homogeneous dose spheres. Final conclusions on the accuracy of the MU calibration and beam modeling are drawn in section 3.4.4.

3.4.1 Lateral profile of a single pencil beam

Although not perfectly describing the lateral profile of a scanned proton beam, treatment planning systems usually use Gaussian distributions to describe the lateral profile since it is computationally efficient and therefore suitable for routine treatment planning. Pedroni et al. [2005] proposed to parametrize the beam width by a so-called double Gaussian. The first, narrow Gaussian component describes the primary beam. The second Gaussian component, which is much broader and flatter, accounts for the low-dose envelope due to large-angle scattering and inelastic collisions [Schwaab et al., 2011, Parodi et al., 2013].

In order to investigate the accuracy of a double Gaussian description, the lateral profiles of a 180 MeV single scanned proton beam, as exported from XiO and from a FLUKA simulation are analyzed. Therefore, the function

$$D(E,z,x) = n(E,z) \times \left(\frac{1 - w(E,z)}{2\pi\sigma_1^2(E,z)} e^{-\frac{x^2}{2\sigma_1^2(E,z)}} + \frac{w(E,z)}{2\pi\sigma_2^2(E,z)} e^{-\frac{x^2}{2\sigma_2^2(E,z)}} \right) \quad (3.9)$$

was fitted to the data points from both XiO and FLUKA, using a MATLAB (The Mathworks Inc., USA) routine provided by Schwaab et al. [2011], Parodi et al. [2013]. $n(E,z)$ is a normalization factor, $(1 - w(E,z))$ and $w(E,z)$ are the relative weights of the narrow and the broad Gaussian with their widths σ_1 and σ_2, respectively, and x is the lateral distance to the beam's main axis. Since the spots in our beam model are not rotationally symmetric, only the central x-z-slice was chosen for the evaluation, for both TPS and MC.

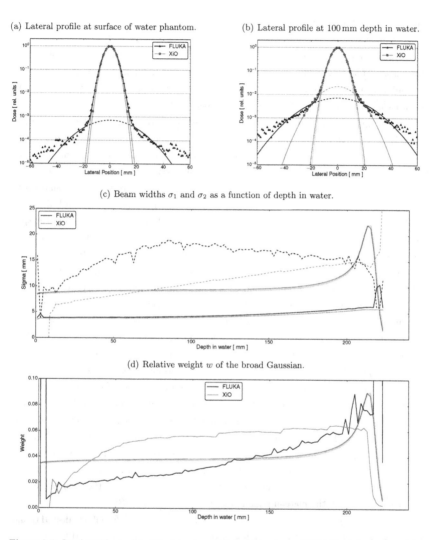

(a) Lateral profile at surface of water phantom.

(b) Lateral profile at 100 mm depth in water.

(c) Beam widths σ_1 and σ_2 as a function of depth in water.

(d) Relative weight w of the broad Gaussian.

Figure 3.5: Semilogarithmic plot of the lateral profiles of a proton beam (180 MeV) on the surface of the water phantom (a) and in the plateau region, at 100 mm depth in water (b). The symbols are the data points extracted from MC simulation (dark gray triangles) and the TPS (light gray squares). The solid lines are the fitted results according to eq. (3.9), the dotted and the dashed lines correspond to the fitted narrow and the broad Gaussian, respectively.

The fitted widths σ_1 (thin solid lines) and σ_2 (dashed lines) of the two Gaussians as a function of depth in water are shown in (c). At low penetration depths, the fitting routine was not able to find solutions for σ_2 of the TPS.

(d) shows the relative weight of the broad Gaussian, according to eq. (3.9). Again, MC and TPS are plotted in dark gray and light gray, respectively.

The dotted black and thick gray lines in (c) and (d) correspond to the dose from the normalization of the fit $n(E, z)$ (cf eq. (3.9)) for MC and TPS, respectively.

However, the other main axis of the elliptical spot can be studied in an analog way and yields similar results.

In figure 3.5 (top), the lateral beam profile for the 180 MeV pencil beam is plotted at the surface of a water phantom (a) and in the plateau of the pencil beam at 100 mm (b). A semi-logarithmic scale is used and the dose heights are normalized to unity in order to show clearly the contribution of the second, broad Gaussian.

On the surface of the water phantom, the profile extracted from the TPS is parametrized by only a single Gaussian, while the contribution of large-angle scattered primaries and secondaries in air and beamline elements is neglected. The simulation confirms, that the halo is more than three orders of magnitude smaller than the central axis dose. Also, at larger distances (± 60 mm) from the spot center, the doses are too low to sum up to relevant contributions to the total dose, even for an extended field. The small difference in the fitted σ_1 can be explained by the fact, that in the TPS model assumption, the spot ellipsis is not tilted [Ele, 2012], whereas for the input spot size of the FLUKA simulation, the width of the main axes of the tilted ellipses were taken [Englbrecht, 2014].

At an intermediate depth at 100 mm in water, the data points extracted from the TPS can be well described by a double Gaussian distribution. The low-dose envelope here is one order of magnitude larger than at the surface and thus can add up to a significant amount of the total dose, if larger field sizes are considered. However, the comparison to the results from FLUKA calculations shows large discrepancies. The width of the XiO's broad Gaussian is significantly smaller than the fit to the simulated data, where the low-dose envelope extends over several centimeters. Furthermore, non-zero entries in the dose exported from XiO can only be found between -25 mm to 25 mm. At larger distances to the beam center, the predicted dose is therefore much smaller than the simulated dose.

The fitted widths σ_1 and σ_2 of the narrow and broad Gaussian, respectively, are plotted for both MC and TPS, against depth in figure 3.5 (c). The depth dependence of the relative width w of the broad Gaussian is shown in (d). It can be seen that the maximum of the low-dose envelope is at an intermediate depth of the scanned proton beam. This is consistent with the results published by [Sawakuchi et al., 2010b] and stresses the influence of the low-dose envelope in the plateau region of scanned fields and SOBPs.

Compared to the relative weights of the broad Gaussian in Schwaab et al. [2011] and Parodi et al. [2013], the relative weights in this study are quite small (< 0.1, neglecting obvious outliers due to the fitting). In these two publications, the measured and simulated doses were integrated over one direction and scored with radial binning, respectively. Comparable weights (0.10 to 0.15) could be found when fitting eq. (3.9) to data sets, but integrated over one axis instead of fitting to the central slices.

Especially a intermediate depths, the comparison of the lateral profile of a scanned proton beam between XiO and MC simulation showed that differences can be seen at larger distances from the beam's central axis. In fact, it is essential to study the influence of these differences when a large number of single pencil beams sum up to an extended field.

3.4.2 Field size dependence of the output factor for monoenergetic square fields

As it has already been mentioned, the impact of the low-dose envelope will be more signifi-cant if several individual pencil beams are superimposed. It is thus reasonable to compare the dose at the center of monoenergetic quadratic layers of different field size. We investi-gated the dependence of the dose output on field sizes between $2 \times 2\,\mathrm{cm}^2$ and $10 \times 10\,\mathrm{cm}^2$ and compared it among TPS calculation, FLUKA MC simulation and measurements. The comparison also includes TPS and MC values for field sizes up to $13 \times 13\,\mathrm{cm}^2$.

3.4.2.1 Methods

The dose distribution for all monoenergetic square fields at a beam energy of 180 MeV, with lateral sizes ranging from 2 cm to 13 cm were calculated with the treatment planning system XiO in a water phantom, with its proximal surface 150 mm prior to the treatment isocenter. The lateral spacing between two adjacent spots in the isocenter plane was 2.5 mm and hence much smaller than the FWHM of the individual pencil beams in the isocenter plane[5]. This ensures that the pencil beams overlap forming a homogeneous dose region in the center of the square fields. The weight of all spots was kept the same for all fields, so no optimization by the TPS was performed. The calculated dose was then exported in DICOM format[6] in order to allow for comparison to the measured and simulated data.

The nine square fields up to 10 cm size were then irradiated in a water phantom, po-sitioned as described above and as calculated in the TPS. A small cylindrical IC (type 31010, PTW-Freiburg, Germany) with radius 2.75 mm and length 6.5 mm, totaling a sen-sitive volume of $0.125\,\mathrm{cm}^3$, was used to measure the dose on the central axis. The main axis of the chamber was placed perpendicular to the beam direction, thus parallel to the scanned field. The dosimetric uncertainty of the IC is 1.3 % in this experiment. Since dose measurements in the BP region are not robust against small depth variations, doses are measured in the plateau region at 100 mm depth, which is about half of the proton range in water. Thus, the positioning uncertainty of the IC in depth, as well as in transversal direction should have a negligible impact on the measured result, since a flat dose region can be assumed around the central axis, due to the used spot spacing. In fact, from MC simulations the dose error due to positioning errors of 0.5 mm could be estimated to be smaller than 0.1 %. For the comparison with the TPS calculated and the MC simulated doses, the WEPL of the entrance window of the IC is taken into account.

---

[5]The FWHMs of the spots in air at the isocenter plane are about $\mathrm{FWHM}_x = 10.0\,\mathrm{mm}$ and $\mathrm{FWHM}_y = 8.5\,\mathrm{mm}$ in x- and y-direction, respectively [Weick-Kleemann, 2013].

[6]DICOM (Digital imaging and communication in medicine) is a common file format for the storage of medical data.

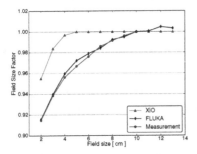 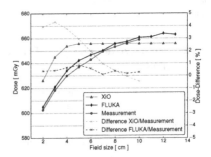

Figure 3.6: Field size dependence of the output factor for monoenergetic square fields. The left plot shows the field size factor (FSF) of the predicted dose (triangles), the measured dose (squares) and the simulated dose (diamonds) for 180 MeV at 100 mm depth in water. On the right, predicted, simulated and measured doses at 100 mm depth in water are plotted against field size (left axis), as well as the relative difference between predicted and measured dose (light gray, dashed line) and simulated and measured dose (dark gray, dashed line), respectively (right axis).

The plans were then simulated with FLUKA, where dose was scored along the central beam axis by a cylindrical USRBIN with a diameter of 5.5 mm and a depth resolution of 0.5 mm. Thus, the scoring size is comparable to the length of the IC's sensitive volume. The statistical uncertainty of the simulation ($< 0.2\,\%$) is small compared to the expected uncertainty of the MU calibration and the dosimetric uncertainty.

3.4.2.2   Results and discussion

Field size factor The field size factor (FSF) was defined as the dose at the center of a given square field with a certain size f, divided by the dose at the center of a square reference field with a size $f = 10$ cm [Sawakuchi et al., 2010b]. As expected, since the contribution of the low-dose envelope increases with a larger number of spots, the FSF decreases with decreasing field sizes (figure 3.6, left). Comparing the measured doses in the center of the 10 cm and 2 cm square fields, results in a FSF of 0.915. This is in perfect agreement with the FSFs deduced from FLUKA simulations. For all measured and simulated field sizes, FSFs agree within 0.6 %. The FSF predicted by the TPS is close to the measured and the simulated FSF for large field sizes (> 7 cm). Still, for smaller field sizes, XiO overestimates the FSF by almost 5 % as compared to measurements.

Sawakuchi et al. [2010b] determined the FSF for 148.8 MeV and 221.8 MeV at slightly greater depths. Nevertheless, their FSFs are similar to our values for fields larger than 4 cm. For smaller fields, their FSFs are noticeably smaller. These differences for small fields can be explained when comparing the spot sizes of their facility to our spot sizes. The reported spot sizes of their facility at the entrance of the water phantom are 12.4 mm and 17.6 mm for the higher and the lower beam energy, respectively. According to the semi-empirical formula presented in Sawakuchi et al. [2010b], a 180 MeV proton pencil beam at their

Table 3.1: Field size dependence of the output factor for a 180 MeV square field. The measured D_{meas}, predicted D_{XiO} and simulated dose D_{MC} at 100 mm depth in water are given for different field sizes. Moreover, the relative differences between the predicted and the measured dose, Δ_{meas}^{XiO}, as well as the differences between the simulated and measured dose, Δ_{meas}^{MC}, are given. The uncertainty of the measured doses is 1.3 %.

Size [cm]	D_{meas} [mGy]	D_{XiO} [mGy]	Δ_{meas}^{XiO}	D_{MC} [mGy]	Δ_{meas}^{MC}
2	602.6	626.3	+3.93%	605.0	+0.40%
3	618.4	645.2	+4.34%	621.0	+0.42%
4	630.0	653.9	+3.80%	634.1	+0.65%
5	637.0	655.9	+2.96%	642.4	+0.84%
6	643.1	655.9	+1.99%	646.9	+0.59%
7	649.6	655.9	+0.97%	650.3	+0.10%
8	653.3	655.9	+0.40%	655.8	+0.39%
9	656.3	655.9	−0.05%	657.5	+0.19%
10	658.9	655.9	−0.46%	660.8	+0.29%
11	-	655.9	-	661.2	-
12	-	655.9	-	663.9	-
13	-	655.9	-	663.0	-

treatment site has a FWHM of 14.9 mm. Hence, our spot sizes in air (FWHM$_x$ = 10.0 mm and FWHM$_y$ = 8.5 mm) are significantly smaller than their spot sizes. That means that at their facility, direct dose deposition of nearby pencil beams plays a more important role for small fields, despite the more narrow spot spacing used in our study.

Grevillot et al. [2011] used the same spot spacing as in our experiments and an ionization chamber with a comparable sensitive volume. They also compared the measured FSFs with the values obtained from GATE MC simulations and could achieve satisfactory agreement within 2 % for all but two points for 4 different beam energies in up to 3 measurement positions. Our measured FSFs are in good agreement with the FSFs at 100 mm depth for the 180 MeV square fields in Grevillot et al. [2011].

However, none of these studies compared measured and simulated FSFs to values predicted by treatment planning systems.

Absolute dose at 100 mm depth In table 3.1 the measured, predicted and simulated doses at 100 mm depth in water are summarized for the different square field sizes. Good agreement can be found between the measured dose and the dose from MC simulations, although the simulated doses are slightly higher for all field sizes. The mean difference is 0.43 % with the largest difference of 0.84 % at a field size of 5 cm. In order to judge whether this small deviation is a systematic error in the MU calibration, further measurements at a several depths and more energies would be necessary. However, the agreement is within the estimated uncertainty of the IC measurement itself.

In contrast, the comparison to the dose values at 100 mm, predicted by XiO clearly shows deviations for small fields. About 3 % difference and more could be observed for fields smaller than 6 cm. Only at field sizes larger than 7 cm, the predicted dose agrees

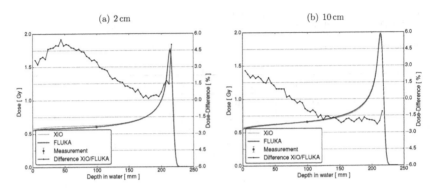

Figure 3.7: Depth dose profiles along the central axis for a 180 MeV, 2 cm (a) and a 10 cm (b) square field. Predicted dose profiles are plotted in light gray and the simulated profiles are plotted in dark gray. The measures dose is illustrated by the circle. The left axes are the corresponding doses in mGy. The black lines denote the relative differences between predicted and simulated dose on the corresponding right axis.

with the measured and simulated doses within 0.5 %. In figure 3.6 (right), the doses and the relative differences are shown for the studied field sizes.

Depth dose profiles   Dose differences due to different field sizes are certainly of interest for the complete penetration depth and not only for one measurement position. Due to the lack of measured data, the predicted dose from the TPS can only be compared to dose output of FLUKA. Figure 3.7 shows two depth dose profiles along the central axis for a 2 cm and a 10 cm monoenergetic square field of 180 MeV.

For small field sizes, XiO overestimates the dose in the plateau by up to 5.5 % as compared to simulation data. Differences become smaller in the vicinity of the Bragg Peak. With increasing field sizes, the differences in the depth dose profiles become smaller. At field sizes of 8×8 cm^2 and larger, TPS and FLUKA agree within 2 %, except close to the phantom surface.

Comparison XiO/FLUKA for a larger energy range   For a field size of 8×8 cm^2, the difference between TPS and simulation was within 2 % for almost all evaluated depths and at the measurement position at 100 mm, the agreement of the measured, the simulated and the predicted dose is within 0.5 %. Therefore this field size was chosen to validate the MU calibration for a larger range of beam energies. The choice of this field size is reasonable, since for the MU calibration of XiO, a field size of 8×8 cm^2 or 10×10 cm^2 is recommended by the vendor [Ele, 2012]. In this part, only simulation and predicted dose profiles are compared since no measured data was available. Energies were ranging from 90 MeV up to 230 MeV in 10 MeV-steps.

In more than 94 % of all points in the depth dose profiles along the central axis, XiO and FLUKA agreed within 3 % and the largest deviations could be observed around the BP, where small misalignments between the two data sets and binning effects are the reason for the large differences. The integral dose over the complete depth profile agrees within 1 % for all evaluated energies. The mean difference between the simulated and the predicted depth dose profiles for all energies is 0.1 % and therefore negligible. Only the slope of both profiles is different, as it could already be seen in figure 3.7. This is due to the obvious differences between TPS and MC in their way of calculating dose.

3.4.2.3   Conclusion

Measurements and detailed MC simulations indicate, that there is a strong dependence of the output factor on the field size. The absolute dose in the center of a square field varies by more than 8 % in the plateau region, when comparing a 2 cm to a 10 cm square field. Especially for small fields the treatment planning system XiO underestimates the influence of the field size on the absolute dose output. In clinical cases implying such small fields, it might therefore be recommendable to compare the predicted dose to measurements or detailed MC simulations.

Still, further investigations are necessary, including measurements in a broader energy spectrum and at more positions. Additionally, in the BP region measurements would be helpful, although care has to be taken there because of the large effect of positioning errors of the IC and the energy dependence of the stopping power ratio for translating dose in air to dose to water.

3.4.3   Model validation for dose to spherical target volumes

Typical treatment plans consist of more than just one single energy layer. Thus, it makes sense to study the influence of the size of the irradiated *volume* on the dose output. Instead of a dose cube as in section 3.3.2, spheres with homogeneous dose and different sizes have been planned in a water phantom to compare the predicted dose to measured and simulated doses.

3.4.3.1   Methods

The dose spheres with diameters ranging from 2 cm up to 10 cm are planned in a water phantom such that their center is always located in the treatment isocenter. The water phantom starts 150 mm upstream of the isocenter. The plans were optimized in order to achieve a dose distribution which is as homogeneous as possible close to the central axis. A spot spacing of 2.5 mm was used for spheres with diameters up to 4 cm and 4.0 mm was used for the larger spheres. The total number of energy layers ranged from 10 (140.13 MeV to 151.97 MeV) for the smallest dose sphere up to 29 (117.90 MeV to 172.39 MeV) for the 10 cm dose sphere.

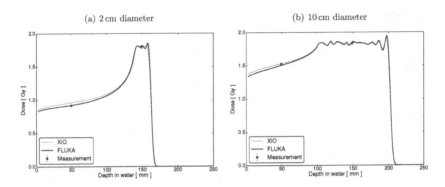

Figure 3.8: Depth dose profiles along central axis for two spherical target volumes with 2 cm (a) and a 10 cm (b) diameter. Predicted dose profiles are plotted in light gray and the simulated profiles are plotted in dark gray. Measured doses are shown by circles.

The water phantom was irradiated according to the treatment plans and dose was measured at the central axis in the plateau at 50 cm depth, as well as in the center of the dose sphere, at 150 cm. The IC used for measurements is the same as described in section 3.4.2.

Simulation and dose scoring were also performed in the same way as in section 3.4.2.

3.4.3.2 Results and Discussion

A summary of the resulting doses and their relative differences is given in table 3.2. In the peak region at 150 mm, measured, predicted and simulated dose nicely agree within 1 % for all sphere sizes. The largest differences are 0.83 % between the XiO and the measured dose and 0.56 % between the simulated and the measured dose. In figure 3.8, a somewhat larger deviation between FLUKA and TPS can be seen in the ripples of the SOBP. This can easily be explained by the coarse grid of 2.6 mm that has been used in TPS dose calculation.

Also, the size of the spheres only marginally influences the dose in the peak region. The largest difference from the mean dose is about 1.2 %.

In contrast, large differences can be seen in the plateau. Like for small square fields, the dose computed by the TPS is significantly higher than the measured dose. With larger sphere sizes, the predicted dose approaches the measured dose until there is practically no more difference for the 10 cm sphere.

Slightly larger differences in the plateau dose can be seen between the simulation and measurement. FLUKA underestimates the plateau dose by up to 1.47 % as compared to measurement. However, the difference does not depend on the sphere size but remains rather constant except for the smallest sphere. Since this means that the FLUKA modeling

Table 3.2: The measured D_{meas}, predicted D_{XiO} and simulated dose D_{MC} at 50 mm and 150 mm depth in water are given for different dose sphere diameters. Δ_{meas}^{XiO} and Δ_{meas}^{MC} are the relative differences between predicted and measured, as well as simulated and measured dose, respectively. The uncertainty of the dose measurements is 1.3 %.

Measurement Position	Diameter [cm]	D_{meas} [Gy]	D_{XiO} [Gy]	Δ_{meas}^{XiO}	D_{MC} [Gy]	Δ_{meas}^{MC}
50 mm	2	0.919	0.962	4.68%	0.922	0.33%
(plateau)	3	1.081	1.103	2.07%	1.065	−1.45%
	4	1.166	1.192	2.21%	1.158	−0.71%
	5	1.247	1.268	1.68%	1.230	−1.37%
	6	1.313	1.329	1.25%	1.294	−1.41%
	7	1.381	1.392	0.77%	1.362	−1.40%
	8	1.435	1.447	0.85%	1.420	−1.04%
	9	1.482	1.490	0.55%	1.460	−1.47%
	10	1.525	1.530	0.30%	1.504	−1.41%
150 mm	2	1.796	1.810	0.76%	1.804	0.43%
(peak)	3	1.838	1.853	0.83%	1.847	0.50%
	4	1.841	1.854	0.70%	1.849	0.43%
	5	1.805	1.820	0.83%	1.810	0.27%
	6	1.820	1.828	0.42%	1.829	0.48%
	7	1.814	1.814	0.01%	1.824	0.56%
	8	1.821	1.821	−0.02%	1.827	0.31%
	9	1.829	1.821	−0.44%	1.833	0.21%
	10	1.839	1.827	−0.65%	1.841	0.11%

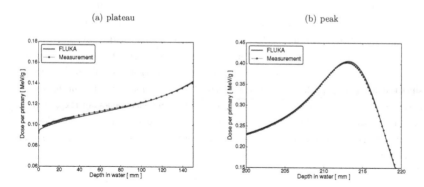

Figure 3.9: Plateau (a) and peak (b) region of the depth dose profile of a 180 MeV single pencil beam. The dose measured with a BP chamber (type 34070, PTW-Freiburg, Germany) is shown by circles. It has been normalized to the area of the dose calculated by FLUKA (solid dark line), using the same scoring radius as the BP chamber. Doses to water are given per primary beam particle. A difference in the peak-to-plateau ratio between measurement and simulation is visible.

of the low-dose envelope works rather well, the reason for this discrepancies has to be a different one. We therefore compared the measured laterally integrated depth dose profile of a single pencil beam to the simulation of the same beam.

In figure 3.9, the plateau and the peak region of the depth dose profile of a 180 MeV single pencil beam can be seen. The relative dose was measured in a water phantom using a plane-parallel BP chamber (type 34070, PTW-Freiburg, Germany). The single pencil beam was simulated in water and dose was scored cylindrically using the radius of the BP chamber's sensitive volume, $r = 4.08$ cm. The measured relative dose was normalized by area to the simulated dose. It can be seen, that also for the single pencil beam, the peak-to-plateau ratio is larger for the simulation than in the measurement. This can also be seen for pencil beams at different energies, though not shown here. However, it has to be stressed that the measurement of the BP chamber is more uncertain in the peak region. This is due to the energy dependence of the stopping power ratio and possible saturation effects of the chamber.

3.4.3.3 Conclusion

Overall good agreement between simulation and measured doses could be achieved for all sphere sizes. The slightly larger differences in the plateau are consistent with independent measurements and simulations of single pencil beams in water. Most likely, small inaccuracies in the physics models of FLUKA and the uncertainties in the measurement mentioned before are the reasons for this difference. Furthermore, in the simulation geometry we did not account for the exact BP chamber geometry but modeled it as water. This however, should only have a small influence on the simulated depth dose profiles [Kurz et al., 2012].

It has to be stressed though, that differences of this order are clinically acceptable, especially when considering that the differences were small even for such complex plans like the large spheres with up to 11 000 individual spots and 29 energy layers.

Differences to the TPS for the small spheres indicate once again, that for small volumes the predicted dose is not accurate and the dose distribution should be validated. However, since the doses in the peak region agree within 0.83 % and the measured dose in the plateau is smaller than the planned dose, no severe changes in the effectiveness of the treatment have to be suspected. Any discrepancies are hence in the safe direction i.e., the dose delivered to healthy tissue could be much lower than tolerance doses set in treatment plan optimization.

3.4.4 Final discussion on the accuracy of the MU calibration

The initial question, whether the low-dose envelope can be responsible for the discrepancies discussed in section 3.3.2 could at least partially be affirmed. The lateral dimension of the SOBP planned there was only 6×6 cm^2. According to the studies performed in this section, the difference in dose between FLUKA and TPS are about 1.5 % for a 180 MeV square field of this size. Assuming that the difference is similar for the lower energies used to form the SOBP (143.75 MeV - 174.33 MeV), the low-dose envelope could indeed sum up to this difference in dose. But to reliably proof this hypothesis, further analyses and measurements for the corresponding beam energies would be necessary which would, however, go beyond the scope of this thesis.

The simulation of a comparable volumetric plan, a sphere with 6 cm diameter, yielded similar results. The dose predicted by XiO agrees perfectly with the simulated dose in the center of the sphere. In the plateau at 50 mm depth in water, the predicted dose is 2.7 % larger than the simulated dose. The findings are therefore consistent and can be explained by the different lateral modeling of a pencil beam by the TPS.

The mean point-to-point dose difference of the MC proton pencil beam scanning model described in Grevillot et al. [2011] was below 2 % in most cases and therefore described as clinically acceptable. The differences between measurements and simulations obtained in our study are even smaller (< 1.5 %) and can therefore without any doubt be regarded as clinically acceptable.

These results confirm, that the approach described in section 3.3.2 is suitable to correctly convert Monitor Units to absolute proton numbers. Assuming an uncertainty of 1.5 % for the conversion is therefore a conservative, but reasonable estimate which will be used for analyses in the following chapter.

4 PET activation studies

The main purpose of this work was to quantify thick-target yields of positron-emitters induced by proton irradiation in the therapeutic energy range. This is the necessary first step towards a clinical implementation of offline PET monitoring. Although experimental cross section values for quantification of proton-induced production of positron emitters already exist, large differences of up to several tens of mb are reported among different data sets [Beebe-Wang et al., 2003, Parodi et al., 2005]. It is therefore suggested to validate cross-sections against measured activation data, taken at the specific PET scanner and beamline that shall be used for PET-monitoring [Seravalli et al., 2012].

A further aspect which makes activation experiments at the RPTC interesting is the proton accelerator used. In contrast to synchrotrons, the beam characteristics of cyclotrons, namely the continuous beam extraction instead of a repetition of beam spills and pauses, makes the assessment of cross-sections more reliable.

However, the accuracy of the measured results strongly depends on the performance of the PET scanner. This study therefore also includes an evaluation of the scanner performance and checks its usability for offline PET monitoring.

This chapter starts with an overview of the materials and methods used for experiments, simulations and data evaluation (section 4.1). This is followed by the results. In section 4.2.1 and section 4.2.2, the integral and depth dependent absolute isotope yields are presented, respectively. Section 4.2.3 shows the findings concerning the lateral activity profiles. In section 4.3 the scanner performance at high activity concentration levels is investigated. Finally, the major uncertainties of these experimental studies are discussed in section 4.4.

4.1 Materials and Methods

In all experiments, dedicated phantoms were β^+-activated by a clinical proton beam and the activity was then measured in list mode in the nearby PET/CT scanner. From this measured data, the proton-induced positron emitter yields were extracted using least-square fitting methods and a mathematical modeling of the activity build-up and decay. The irradiation was also simulated in FLUKA using the updated beam model described in chapter 3. To evaluate the performances of FLUKA's internal models for hadronic fragmentation, as well as for validating the experimental cross-sections used for the calculation of β^+-emitter yields, simulated data had to be processed to make them comparable to experimental data.

The first section therefore provides an overview of the selected phantoms and describes their physical properties. In section 4.1.2, the experimental setup is explained, including both irradiation and PET measurement. This is followed by an outline of the simulation setup in section 4.1.3. Finally, in section 4.1.4 the methods used for data evaluation are explained.

4.1.1 Phantoms

About 98% of the atoms in human tissue are hydrogen, oxygen and carbon atoms [Latscha et al., 2011]. Obviously, hydrogen nuclei can not fragmentate yielding positron emitters. Therefore, oxygen-16 and carbon-12 (^{16}O and ^{12}C), totaling about one third of the isotopic abundance in human tissue, are the most promising candidates as reaction partners for positron emitter production reactions. Because of the unavoidable time delay between irradiation and PET-measurement in the adopted offline imaging concept, only positron-emitting isotopes with half-lives in the order of some minutes are expected to contribute to the measured activity signal. Thus, the relevant isotopes are ^{11}C, ^{13}N and ^{15}O with half-lives of 1220.0 s, 597.9 s and 122.2 s, respectively [nud]. Despite the half-life of 70.6 s, which is still large enough to allow a measurable signal within the first minutes of data acquisition, ^{14}O was not included in the data analysis, since the production cross-section is more than one order of magnitude smaller than the production cross-section of ^{15}O [Otuka et al., 2014]. For ^{11}C, the production channels are the ^{12}C(p,pn)^{11}C reaction, as well as the oxygen channel, ^{16}O(p,3p3n)^{11}C. ^{13}N and ^{15}O are produced in the two proton-induced reactions ^{16}O(p,2p2n)^{13}N and ^{16}O(p,pn)^{15}O, respectively.

4.1.1.1 Phantom choice

The choice of the actual phantom materials for the activation experiments was inspired by Bauer et al. [2013]. The investigation of the production channels requires homogeneous phantom materials with known chemical composition. In order to retrieve a signal that can be clearly attributed to one, or only few production channels, the ideal phantom consists of only one element in total or one element besides hydrogen. For more complex atomic compositions, the observed activities are the sum of the contributions of all different positron emitters and it becomes therefore more difficult to distinguish between the individual contributions. Candidates for the study of the carbon production channel of ^{11}C would be graphite and polyethylene (PE, C_2H_4). Bauer et al. [2013] reported some issues with the graphite phantom. The ionization potential of this material, which is essential for correct simulations, could not be determined correctly because of its anisotropic structure. Therefore, only the latter material, PE, was used for the ^{12}C(p,pn)^{11}C channel.

Table 4.1: Overview of the phantoms and their sizes used in the subsequent experiments. Their chemical compositions are given with their densities and ionization potentials, as they are used in the simulation input.

Material	Size [cm^3]	Density [g/cm^3]	I_p [eV]
PE (C$_2$H$_4$)	$10 \times 10 \times 35$	0.959	56.0
Gelatine (H$_{66.2}$O$_{33.1}$C$_{0.7}$)	$10 \times 10 \times 35$	1.00	73.0
PMMA container (C$_5$H$_8$O$_2$)	$12 \times 11 \times 38^1$	1.184	71.0
PMMA (C$_5$H$_8$O$_2$)	$10 \times 10 \times 35$	1.184	71.0

For the study of the oxygen channels, water (H$_2$O) would be the best phantom material if only the chemical composition is regarded. But because of diffusion and motion due to the phantom transport, liquid water can not be used for the investigation of the spatial distribution of produced β^+-emitters. To remedy this disadvantage, gelatine serves as replacement for liquid water. Hence, 3.5 l of distilled water were boiled with 110 g of Agar-Agar (an algae-based gelling agent) and filled into dedicated PMMA boxes where the gelatine solidifies. According to Sommerer et al. [2009], the contamination due to the addition of gelatine powder to water is negligible and changes the chemical composition to H$_{66.2}$O$_{33.1}$C$_{0.7}$.

The third phantom material used in this work is a block made of PMMA (C$_5$H$_8$O$_2$). This material is commonly used in clinical practice for quality assurance measurements. Due to its chemical composition, all of the four investigated production channels can be found and the material is therefore meant as consistency check for the derived isotope yields of the other phantom materials.

In table 4.1 all the phantom materials are listed together with their geometrical dimensions, densities and ionization potentials. The determination and the importance of the latter two properties is described in the following subsection.

4.1.1.2 Determination of the phantom properties

The beam range in a medium is mainly determined by the medium density and its ionization potential. Thus, accurate knowledge of these parameters is fundamental for correct simulation of the range in the phantoms. Tabulated values of ionization potentials have generally large uncertainties and material densities may vary for different fabricates. Therefore, it is favored to determine them experimentally for the phantoms used in this study.

The density of PE and PMMA could be obtained in a straight-forward way by accurately measuring the phantom dimensions and weights. This was not possible for the gelatine, since filling the hot and liquid gelatine into the PMMA box deforms the containers' walls by up to 1 mm on each long side. That means, one can not draw conclusions for the gelatine dimensions from measuring the inner dimensions of the empty boxes. But also measuring the gelatine's dimensions outside the box is not possible due to its flabby consis-

---

[1]outer dimensions

(a)   (b)

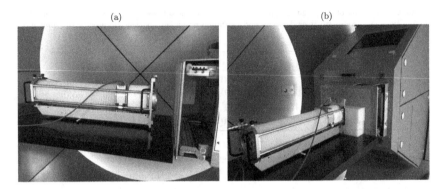

Figure 4.1: Setup for the WEPL determination of the phantoms, (a) without any phantom material in the beam, (b) with one PE phantom in the beam.

tency. Therefore, the density of water at room-temperature is used as density of gelatine. Comparing measured and simulated proton ranges in water and gelatine confirms that for the chosen density and ionization potential, measured ranges could be reproduced by simulations within 0.3 mm.

The ionization potentials of the three materials can not be measured directly. Thus, the WEPL of the materials were measured and the ionization potentials were deduced as described in Kurz et al. [2012]. The setup for these measurements can be seen in figure 4.1: the phantoms were placed with their short side along the beam direction between the beam nozzle and the water column (figure 4.1 b). A water column equipped with a movable BP chamber (PTW Peakfinder, modified by ACCEL, now VARIAN) and a fixed reference ionization chamber at its entrance was used to record the depth dose profiles for two different beam energies (158.53 MeV and 178.01 MeV) for each phantom material. Two further depth dose profiles were acquired in absence of any phantom material. For the gelatine phantom, two depth-dose-profiles were recorded for the filled boxes and two additional profiles for the empty boxes.

A fourth-order polynomial was then fitted to all measured data sets including only dose values larger than 75 % of the maximum dose value, in order to precisely determine the Bragg peak position. The shift of the Bragg peak position with and without phantom material determines the WEPL of these materials. In an iterative trial-and-error process the ionization potential of the materials was adjusted in the FLUKA input parameters (`MAT-PROP` card) until the measured Bragg peak shifts could be reproduced by the simulations within < 0.1 mm for PE, PMMA and the PMMA container and within < 0.3 mm for gelatine.

(a) Irradiation   (b) PET measurement

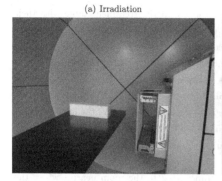 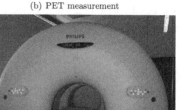

Figure 4.2: Experimental setup of the activation experiments. Using the laser alignment system, the phantom (here: PE) is positioned on the treatment table such that its geometrical center is located in the treatment isocenter (a). The beam is coming from the right.
After irradiation, the phantom (here: gelatine) is placed in the PET/CT scanner with its main axis perpendicular to the scanner's main axis (b).

Table 4.2: Summary of the phantom irradiation parameters: Day of the irradiation, phantom type, total number of delivered protons (calculated from the recorded MUs as described in section 3.3), duration of the irradiation, time delay between irradiation end and start of the PET acquisition, and PET acquisition duration. The experimental data sets marked with an asterisk (*) were used for the evaluation of the scanner performance.

Day	Phantom	Energy [MeV]	N_p	t_{irr} [s]	t_{del} [s]	t_{meas} [min]
05/16/14	PE	107.57	1.05×10^{11}	13.9	159.8	36
05/16/14	PE	158.53	1.05×10^{11}	4.2	183.9	36
08/20/14 *	PE	158.53	2.10×10^{11}	8.3	186.1	36
08/20/14 *	PE	158.53	3.15×10^{11}	13.4	176.7	36
05/19/14	Gelatine	126.53	1.05×10^{11}	8.5	207.4	34
05/22/14	Gelatine	178.01	1.11×10^{11}	5.0	176.5	34
09/09/14 *	Gelatine	126.53	3.01×10^{10}	2.2	169.7	36
09/09/14 *	Gelatine	178.01	3.03×10^{10}	1.6	190.8	36
06/26/14	PMMA	126.53	2.11×10^{11}	15.8	180.8	34
06/26/14	PMMA	178.01	1.11×10^{11}	5.1	173.0	34

4.1.2   Experimental setup

4.1.2.1   Irradiation and PET acquisition

For the β^+-activation, the phantoms were positioned on the treatment table with their geometric center in the treatment isocenter, using the laser alignment system. This was done such that the phantoms' long side was coinciding with the beam direction (figure 4.2 a). The phantoms were irradiated with mono-energetic pencil beams along their central axis. The beam energies were chosen to be comparable to the activation experiments performed at HIT [Bauer et al., 2013] and GSI [Hildebrandt, 2012]. While the gelatine and the PMMA phantoms were irradiated with 126.53 MeV and 178.01 MeV, the beam energies for PE were slightly lower (107.57 MeV and 158.53 MeV). The activation experiments were performed on six different mornings after daily QA and usually before patient treatment started.

After the irradiation, the phantoms were transported to the room nearby, where the PET/CT scanner is located. The phantoms were placed on the scanner table according to markings that had been made before irradiation. This allowed accurate positioning while keeping the time delay between irradiation and measurement short. They were placed with the phantoms' long axis perpendicular to the scanner's main axis (figure 4.2 b). Alignment parallel to the scanner's main axis was not possible, since the penetration depth of the beams for the higher energies is larger than the axial FOV of the scanner.

PET acquisition started after the initial scout scan and was followed by a CT scan, necessary for the attenuation and scatter correction of the PET image. The acquisition time for the PET measurement was set to 34 min or 36 min. The total time delay between end of irradiation and start of the PET acquisition was around 3 min and was mainly due to the time required for the initialization of the scanner and the scout scan.

Since accurate knowledge of the irradiation time t_{irr} and the time delay between irradiation end and PET acquisition start, t_{del}, is crucial for correct data analysis, the whole procedure was recorded with a camera. This allowed to determine the relevant times afterwards by the analysis of the captured video, excluding errors due to reaction time. With this method, sub-second accuracy could be achieved.

In table 4.2, the relevant irradiation and acquisition parameters are summarized. The number of delivered particles has been computed from the prescribed MU using the conversion presented in section 3.3.

4.1.2.2   PET image reconstruction

In contrast to *frame mode* acquisition which is commonly used in clinical routine, the PET data of the experiments were acquired in *list mode*. This means, that each individual coincidence event is stored with the location of the corresponding detector pair and time mark in a *.list-file. After completing the acquisition, an offline sorter protocol can be applied to the list mode data file. This protocol allows to bin list mode data into individual sinograms with user-defined frame duration. The sinograms are then saved in binary *.scn-files and can be reconstructed using the scanner's reconstruction program PetView (Philips Healthcare, Eindhoven, The Netherlands). Different time frame schemes were tested for the dynamic reconstruction in order to determine the dependency of the fit results on the selection of time frames. The protocol used for reconstruction of the experimental data is based on a clinical protocol with only minor modifications. It performs two dynamic reconstructions, where the second reconstruction needs the reconstructed image of the first reconstruction as input data. As reconstruction algorithm, the 3D-RAMLA (3D Row Action Maximum Likelihood Algorithm) was chosen and the following corrections were applied to the data:

- **Attenuation Correction.** The additional 120 kVp CT transmission scan acquired after PET measurement was used to calculate the attenuation correction factors for the 511 keV annihilation photons. The scanner internally converts the measured CT Hounsfield units (HU) to the desired attenuation coefficients based on measured HU of cortical bone, using a bilinear conversion curve as described in Bai et al. [2003].

- **Randoms Correction.** As already mentioned in section 2.2.1, random coincidence events due to two unrelated 511 keV-photons from different positron annihilation events raise the background of the image. Therefore, the sinograms were corrected using the option *subtract*. During PET acquisition two data sets are acquired - one normal data set with the standard (prompt) time window that contains both random and true events, and a second data set with a delayed time window which contains only random events. The true counts can then be obtained by subtracting the delayed counts from the prompt counts.

- **Scatter Correction.** While passing through matter, the 511 keV-photons undergo Compton scattering and may be detected by a detector pair (cf section 2.2.1). This results in a blurring and an increased background of the image. In order to correct for the scattered coincidences, the option *Single Scatter Simulation* was selected. A Monte Carlo simulation based on a first reconstructed image, where no scatter correction was applied, estimates the number of scattered events. The scatter data is then subtracted from the emission sinogram.

- **Decay Correction.** When using PET scanners in diagnostics, one radioactive tracer with a given half-life is injected to the patient. The scanner corrects the counts in each frame for the decay of this radionuclide. But since in the activation experiments, more than one radionuclide is present in the phantom and the half-life is the key to distinguish between the different radionuclides, decay correction is not desired here. However, the reconstruction only returns absolute activity concentrations if decay corrections are applied. So, to keep the correction small, ^{22}Na ($T_{1/2} \approx 2.6$ a) was the isotope selected for decay correction. Unlike in Bauer et al. [2013], no β^+-branching ratio correction had to be taken into account.

- **Normalization.** PET scanners have thousands of detectors and the detection efficiency of each detector pair varies from pair to pair. These efficiency variations have to be taken into account and corrected for. To this aim, the option *Efficiency* was chosen.

- **Dead time correction.** An internal correction that should account for the dead time losses of the scanner was activated to correct the measured data.

After reconstruction, the images with absolute activity values in Bq/ml were exported in standard DICOM format allowing for further analysis.

4.1.3   MC calculation of proton induced positron emitter distributions

For the Monte Carlo calculation of positron emitter yields, two independent approaches were used. The first approach is based on the internal phenomenological models of the FLUKA code. In order to get an output file storing the spatial distribution of residual β^+-emitting isotopes, which are produced in nuclear interactions between the transported particles and the tissue nuclei, the user-routines usrini.f, usrrnc.f and usrout.f together with the RESNUCLEi card in the input file are needed (cf section 3.1.2). However, the calculation using the internal models of the currently officially available FLUKA version (2011.2b.6) has been reported not to be the most accurate method for calculating positron emitter yields [Parodi et al., 2007a, Seravalli et al., 2012].

As long as newer releases with improved internal modeling are not available, it is preferable to use experimental cross-sections for the β^+-emitter yield calculation. For this purpose, a modified version of the fluscw.f user-routine (cf section 3.1.2) was used to combine the energy-dependent proton fluence $\Phi(E)$ during runtime with energy-dependent reaction cross-sections $\sigma_{X \rightarrow Y}(E)$. Thus, the amount of β^+-emitters Y produced in a nuclear interaction between a proton and a nucleus X in a scoring bin of volume ΔV, is [Parodi et al., 2007a]:

$$N_Y = \int \frac{\mathrm{d}\Phi(E)}{\mathrm{d}E} \frac{f_X \rho N_A}{A_X} \sigma_{X \rightarrow Y}(E) \Delta V \mathrm{d}E \ . \tag{4.1}$$

In this equation, ρ is the medium density, f_X and A_X are the fraction by weight and the atomic weight of the nucleus X, respectively, and N_A is the Avogadro number.

Using this approach, only positron emitters produced by protons are considered, while the contribution by other secondaries like neutrons is omitted. However, it has been found out that the latter contribution is indeed not significant, especially within the irradiation area [Parodi et al., 2007a].

The cross-section data used in this study are fine-tuned values from the Heidelberg Ion-Beam Therapy Center (HIT) [Bauer et al., 2013], which were based on cross-section data from Parodi et al. [2007b].

4.1.4   Data evaluation

The analysis of the measured and the simulated data sets were performed using self-written Python routines using the packages NumPy, SciPy, pydicom and lmfit-py. The basic functionalities of these routines are presented in this section.

4.1.4.1   Analysis of the measured PET data

In a first step, all DICOM files from one PET/CT acquisition, each corresponding to one 2-D slice of the PET or CT images, were imported by this routine and their pixel data were sorted according to their measurement time frames (PET only) and their spatial coordinates (PET and CT). Both images are then co-registered and the phantom location in the PET image is determined by its location in the CT image. The part of the PET image in which the phantom was placed, together with a margin of a few centimeters around, is now cut out of the entire image and is used for further analysis. Thus, for each reconstructed time frame there is one 3-D array of the measured activity distribution, including only the measured activity at the phantom location and the small margin around.

Since the measured activity concentration in each voxel, $A(\vec{r}, t)$, is the sum of the contribution of the decays from all present β^+-emitting isotopes i, it can be expressed for each measurement time frame according to

$$A(\vec{r}, t) = \sum_i A_{0,i}(\vec{r}) \cdot \exp(-\lambda_i t) \tag{4.2}$$

[Bauer et al., 2013]. In this equation, t is the time between the end of the irradiation and the center of each time frame, $\lambda_i = \ln 2 / T_{1/2,i}$ is the decay constant of isotope i with half-life $T_{1/2}$. For gelatine and PMMA, the expected isotopes i are ^{11}C, ^{13}N and ^{15}O. For PE, only the first radionuclide is expected. ^{10}C is not included in the evaluation because of its short half-life and its lower production cross-section. Hence, the initial activity after irradiation end for a β^+-emitter i in a voxel of interest, $A_{0,i}(\vec{r})$, can be determined by a weighted least square fit to the time-resolved decay curve of the measured activity in this voxel [Bauer et al., 2013]. The fit package lmfit-py performs this least square fit and additionally returns the uncertainty of the fit. In figure 4.3, a 2-D distribution fitted initial activities for the three individual radionuclides are shown for the gelatine phantom after irradiation with a 178.01 MeV pencil beam.

Although the fit can be performed for each voxel individually, it is reasonable to integrate the measured data laterally prior to the fit. This increases the data statistics and therefore improves the quality of the fit results, yielding initial activities for each isotope depending on the penetration depth of the proton beam [Bauer et al., 2013]. By integrating the measured activity concentration over the complete phantom before fitting, the initial proton-induced β^+-activity contribution of each radionuclide can be retrieved for the whole phantom.

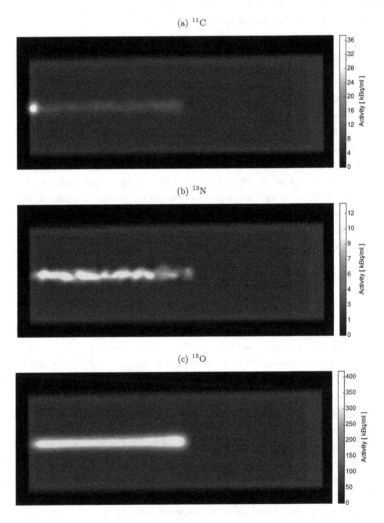

Figure 4.3: Overlay of the fitted initial activities of ^{11}C, ^{13}N and ^{15}O (from top to bottom, respectively) on the phantom CT image (gray-scale) for gelatine irradiated with a 178.01 MeV proton beam.

However, the quantities of interest in this study are the production rate and the production yield of each radionuclide per proton. In this work, the production yield N_i refers to the number of radionuclides of isotope i at the end of irradiation, if not specified differently. The production rate P_i is a quantity describing how many radionuclides of species i are produced per unit time. The duration of the here presented irradiation experiments is rather short. Thus, an error in the time measurement of $0.1\,\mathrm{s}$ would lead to an error in the production rate of several percent. For this reason, final results are given as production yields, which is the quantity less sensitive to uncertainties in the irradiation time measurements.

In Parodi et al. [2002], the calculation of these quantities is performed for the more complex synchrotron irradiation, where the dose is delivered in several spills. For cyclotron irradiation, the dose is delivered in one single spill which makes the calculation easier. Assuming constant beam current, the production rate P_i of isotope i is

$$P_i = \frac{A_{0,i}}{1 - \exp(-\lambda_i t_{\mathrm{irr}})} \, , \tag{4.3}$$

where t_{irr} is the duration of the irradiation. With this equation, the yield of the i-th radionuclide becomes

$$N_i = \frac{P_i}{\lambda_i} \left(1 - \exp(-\lambda_i t_{\mathrm{irr}})\right) = \frac{A_{0,i}}{\lambda_i} \, . \tag{4.4}$$

4.1.4.2   Analysis of the data from FLUKA

The format of the output of FLUKA MC simulations for the spatial distribution of β^+-emitter production yields is identical for both methods described in section 4.1.3. An ASCII file is created for each USRBIN specified in the simulation input file, containing the three-dimensional distributions of the production yields for the selected radionuclides together with their statistical error. As already mentioned in chapter 3, the output is normalized per primary beam particle.

The initial activity distribution at the end of irradiation can then be computed via the simulated production rate [Bauer et al., 2013]

$$P_i(\vec{r}) = \frac{R_i^{\mathrm{MC}}(\vec{r}) \cdot N_p}{t_{irr}} \, , \tag{4.5}$$

where $R_i^{\mathrm{MC}}(\vec{r})$ is the spatial production yield from the MC simulation and N_p is the number of protons. With this equation and again assuming a constant dose delivery, the activity contribution of the i-th radionuclide, right after the irradiation is

$$A_{0,i}(\vec{r}) = G(\vec{r}) * \left[P_i(\vec{r}) \cdot \left(1 - e^{-\lambda_i \cdot t_{\mathrm{irr}}}\right) \right] . \tag{4.6}$$

In order to account for the finite imaging resolution of the PET scanner, a Gaussian convolution kernel $G(\vec{r})$ is applied to the 3-D data array. The standard deviation of the convolution kernel was varied until the proximal slope of the depth-activity profile from the simulation fitted best the measured profile. The evaluation of all acquired profiles suggested a standard deviation of $\sigma = 4.0\,\text{mm}$ [2]. The production yield can then be calculated as for the measured PET data, using eq. (4.4).

For the comparison between the activity calculated with FLUKA and the activity measured in an arbitrary time frame with duration t_{frame}, the average activity concentration in a voxel of volume V_{vox} can be calculated as:

$$\langle A \rangle_i(\vec{r}) = G(\vec{r}) * \left[\frac{1}{V_{\text{vox}}} P_i(\vec{r}) \left(1 - e^{-\lambda_i \cdot t_{\text{irr}}} \right) \frac{e^{-\lambda_i \cdot t_{\text{del}}} \left(1 - e^{-\lambda_i \cdot t_{\text{frame}}} \right)}{\lambda_i \cdot t_{\text{frame}}} \right] . \qquad (4.7)$$

4.2 Results

A very important step towards the validation of the experimental cross-sections is the quantification of the proton-induced positron emitter production yields. The integral isotope yields for the three phantom materials are discussed in section 4.2.1. This is followed by the analysis of the depth dependent isotope yields in section 4.2.2. Since not only the distal, but also the lateral activity profile bears useful information for PET treatment verification, in section 4.2.3 the lateral activity profiles are analyzed.

4.2.1 Integral absolute isotope yields

Two only slightly different approaches were used in this work to obtain the integrated isotope yields. For the first method, the measured activity of each time frame was integrated over the complete phantom. Using the fitting routine described in section 4.1.4, the initial activity contribution of each expected radionuclide and thus the production yield in the complete phantom was computed.

The second approach uses the same fitting routine, but laterally integrated activities serve as input data. This method directly gives the depth-resolved distribution of the production yield which will be discussed in section 4.2.2. Summing up the yields of all depth slices, again results in the total production yield of the complete phantom.

The comparison of the calculated yields from both approaches can be used to check the consistency of the fit results, since both approaches should give the same results. Indeed, no significant differences in the calculated production yields between the two methods could be found. Differences larger than 1 % can only be found for ^{13}N.

A compilation of all fitted β^+-emitter production yields for the entire phantoms is given in table 4.3. In figure 4.4, the measured activity, the total fitted activity and the resolved contribution of each isotope is plotted against time for the PE phantom (a) and

---

[2]Taking a constant standard deviation of the Gaussian convolution kernel for the entire PET image is only an approximation, since the spatial resolution of a PET scanner is better in the center of the FOV than in the edges. Therefore, in the vicinity of the distal fall-off of the measured activity profiles, smaller standard deviations would model the spatial resolution of the PET scanner better.

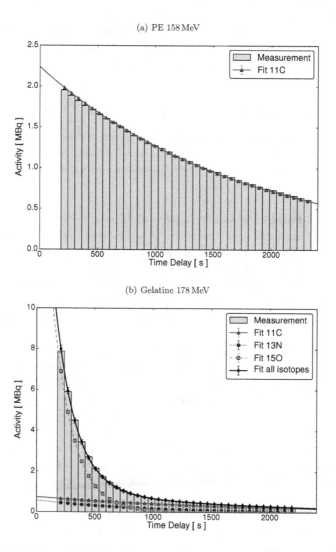

Figure 4.4: Total proton-induced β^+-activity as a function of the time delay from end of irradiation for the PE (a) and gelatine (b) phantoms. The integrated activity measured by the PET scanner at each time frame is displayed as bars. The lines correspond to the fitted activities of the individual isotopes, as denoted in the figures legends. Additionally, the total fitted activity is plotted in black with diamonds in figure (b). Due to dead time losses, the fit exceeds the measured data in the first time frames, while in late time frames, the fit is lower than measurements. This is further discussed in section 4.3.

the gelatine phantom (b) at 158.53 MeV and 178.01 MeV, respectively. Due to dead time losses, the fit exceeds the measured data in the first time frames, while in late time frames, the fit is lower than measurements. This is further discussed in section 4.3.

In section 4.2.1.1, the influence of the time frame choice on the fit results in analyzed, before the measured yields are compared to simulated yields and to previous experimental data in sections 4.2.1.2 and 4.2.1.3, respectively.

4.2.1.1 Influence of the time frame choice

It has been reported, that the choice of the time frames can lead to considerable deviations in the fitted radionuclide contributions when rebinning the list mode data (cf section 4.1.2.2) [Bauer et al., 2013]. Therefore, reconstructions with different time schemes were performed and used for fitting.

Gelatine and PMMA   Because the half-lives of ^{11}C and ^{15}O differ by about a factor of 10, it seems reasonable to use time frames of variable length for the gelatine and the PMMA phantoms, in order to get comparable statistics in each frame. For these hereafter simply referred to as *variable* time frames, the following sequence was chosen: 8×0.5 min + 6×1 min + 6×2 min + 3×4 min. Also, a constant binning scheme was used with 22×1.5 min frames, where the last minute of the 34 min PET acquisition was ignored. For the 178.01 MeV irradiation of gelatine, an additional reconstruction has been done with 34×1 min.

In general, similar findings were obtained as reported in Bauer et al. [2013]. For the high energy irradiation of gelatine, no noticeable effect on the fitted yields could be found when using the two constant time schemes 22×1.5 min and 34×1 min (cf table 4.3). The fitted yields of ^{11}C and ^{15}O agree within 0.4 %. The only larger difference can be seen for ^{13}N, where the resulting yield is about 7 % larger. This is however still within the general fit uncertainty of the ^{13}N yield.

Using variable instead of constant time frames has a larger effect on the results and the fit uncertainties. Differences in fitted isotope yields up to 18.8 % for ^{11}C and even larger for ^{13}N could be observed. Bauer et al. [2013] explained this by the fact that the length of the time frames changes within the period of a half-life. This has a negative effect on the convergence of the fit, which can be seen in the increased fit uncertainties as compared to the fit over constant time frames (cf table 4.3). For gelatine, the fit uncertainty of ^{11}C exceeds 10 % and is about twice as large when using variable instead of constant time frames. For this reason, reconstructions from constant time frames were preferred to variable time frames.

PE   Different from the other considered phantoms, in PE phantoms ^{11}C is the only radionuclide expected to be measured. Because of its relatively long half-life, only constant time frame length was used for binning the list mode data. The two selected schemes were

Table 4.3: Summary of all fitted positron emitter production yields after irradiation for the different phantom materials (PE, GEL, PMMA). The numbers in the first column stand for the beam energies. In the second column, the number of time frames and the frame duration is given. The two methods of fitting - integrate laterally and fit each slice individually (L) and integrate over complete phantom and fit (A), are written in parentheses. In the other columns, the production yields after irradiation of ^{11}C, ^{13}N and ^{15}O are given with the calculated fit uncertainty, respectively. Yields are given per 10^6 delivered protons.

Phantom	Frames/Method	$N(^{11}\text{C})$	ΔN	$N(^{13}\text{N})$	ΔN	$N(^{15}\text{O})$	ΔN
PE107	36 × 1 min (L)	20707	0.48 %	-	-	-	-
	36 × 1 min (A)	20766	0.19 %	-	-	-	-
	24 × 1.5 min (L)	20530	0.53 %	-	-	-	-
	24 × 1.5 min (A)	20567	0.25 %	-	-	-	-
PE158	36 × 1 min (L)	37443	0.45 %	-	-	-	-
	36 × 1 min (A)	37487	0.19 %	-	-	-	-
	24 × 1.5 min (L)	37041	0.48 %	-	-	-	-
	24 × 1.5 min (A)	37073	0.23 %	-	-	-	-
GEL126	variable (L)	6902	10.55 %	3124	21.93 %	22131	1.44 %
	variable (A)	6960	3.85 %	3089	8.22 %	22167	0.53 %
	22 × 1.5 min (L)	7720	5.52 %	2422	18.74 %	21537	1.45 %
	22 × 1.5 min (A)	7782	2.25 %	2398	7.80 %	21544	0.60 %
GEL178	variable (L)	9879	10.99 %	5640	18.12 %	35692	1.19 %
	variable (A)	9883	5.96 %	5654	9.80 %	35703	0.64 %
	34 × 1 min (L)	11632	5.24 %	4380	14.57 %	35169	1.12 %
	34 × 1 min (A)	11716	2.01 %	4340	5.67 %	35177	0.43 %
	22 × 1.5 min (L)	11676	5.27 %	4067	15.91 %	35145	1.16 %
	22 × 1.5 min (A)	11743	2.15 %	4034	6.59 %	35149	0.48 %
PMMA126	variable (L)	21552	3.08 %	1310	44.43 %	7540	2.51 %
	variable (A)	21543	1.23 %	1333	17.33 %	7540	0.98 %
	22 × 1.5 min (L)	22816	1.01 %	332	70.18 %	7218	1.57 %
	22 × 1.5 min (A)	23167	0.00 %	0	0.00 %	7321	0.00 %
PMMA178	variable (L)	34038	4.02 %	2334	50.86 %	12390	3.03 %
	variable (A)	34044	1.73 %	2351	21.86 %	12392	1.30 %
	22 × 1.5 min (L)	36591	1.26 %	346	128.03 %	11864	2.03 %
	22 × 1.5 min (A)	36969	0.60 %	0	0.00 %	11963	0.86 %

24 × 1.5 min and 36 × 1 min frames. As it can be seen in table 4.3, the time frame choice has a minor impact on the fitted yields. The largest deviation is about 1.1 %. Also, fit uncertainties are hardly affected by the time frame choice.

4.2.1.2 Comparison between measured and simulated data

In this subsection, fitted production yields assuming instantaneous production[3], are compared to yields calculated in FLUKA MC simulations for each phantom material. Both the internal hadronic models and experimental cross-sections are included in the comparison. All isotope yields are summarized in table 4.4.

PE In the PE phantom, measurable positron emitters are produced only in the $^{12}C(p,pn)^{11}C$ channel. The MC simulations using the experimental cross-sections are in very good agreement with the measurements. The deviation from the measurements is 3.8 % and 0.8 % for 107.57 MeV and 158.53 MeV, respectively. On the other hand, yields calculated using the FLUKA 2011 internal hadronic model are about 11.7 % and 10.8 % lower than measured yields for the two energies, respectively.

Gelatine Larger differences can be seen for the gelatine phantoms. FLUKA simulation results using external cross-sections systematically underestimate the amount of produced positron emitters. The dominant radionuclide, ^{15}O, is measured to be 8.7 % and 9.7 % more abundant for the lower and the higher beam energy, respectively, than predicted by the simulation using experimental cross-sections. The internal FLUKA models, on the other hand, predicts ^{15}O yields which are 11.4 % for 126.53 MeV and 19.9 % for 178.01 MeV higher than our measurements.

Similar findings with comparable differences could be obtained for the ^{11}C production when comparing fitted yields from the measurement to the FLUKA simulation using experimental cross-sections. The internal FLUKA models overestimate production yields for this radionuclide by almost 8 %.

For the other production channel in the gelatine phantom, $^{16}O(p,2p2n)^{13}N$, differences between the RPTC data and the simulation results are difficult to judge because of the large fit uncertainty of 18.7 % and 14.6 % for 126.53 MeV and 178.01 MeV, respectively. Anyway, both FLUKA simulation approaches underestimate the amount of ^{13}N. The internal models predict yields that are more than 50 % lower than calculated by the fit based on the measurements at RPTC. For the simulation using the external cross-sections, the underestimation reduces to a 24 % lower yield.

---

[3]In order to make fitted yields comparable amongst measurements at various facilities with different beam structure and simulations, instantaneous production of β^+-emitters is assumed. To this aim, production rates calculated according to eq. (4.3) are multiplied by the corresponding beam delivery time.

Table 4.4: Comparison of production yields assuming instantaneous production for all phantom materials and beam energies amongst RPTC, HIT [Bauer], GSI [Hildebrandt] (where available) and the two FLUKA simulation approaches. Yields are per 10^6 delivered protons.

Phantom	Data set	Energy [MeV]	$N(^{11}C)$	$N(^{13}N)$	$N(^{15}O)$
PE	RPTC	107.57	20789	-	-
	FLUKA (int)	107.57	18353	-	-
	FLUKA (exp)	107.57	21569	-	-
PE	RPTC	158.53	37479	-	-
	FLUKA (int)	158.53	33419	-	-
	FLUKA (exp)	158.53	37164	-	-
Gelatine	RPTC	126.53	7739	2434	22060
	HIT	126.58	6860	2412	23733
	GSI	126.31	6777	2763	26643
	FLUKA (int)	126.53	7935	1158	24584
	FLUKA (exp)	126.53	7019	1841	20140
Gelatine	RPTC	178.01	11649	4393	35670
	HIT	178.04	10313	3868	38369
	GSI	178.24	11234	5125	40769
	FLUKA (int)	178.01	12573	2074	42760
	FLUKA (exp)	178.01	10129	3358	32218
PMMA	RPTC	126.53	22919	335	7546
	HIT	126.58	22513	1451	8101
	GSI	126.31	30908	1192	11997
	FLUKA (int)	126.53	20386	461	9768
	FLUKA (exp)	126.53	22620	733	7949
PMMA	RPTC	178.01	36634	347	12036
	HIT	178.04	36085	1564	14030
	GSI	178.24	58535	2041	20707
	FLUKA (int)	178.01	34289	794	16512
	FLUKA (exp)	178.01	35811	1298	12317

PMMA In terms of ^{15}O production yield, the two phantom materials gelatine and PMMA are comparable. Normalized to their ^{16}O target nuclei density[4], the experimental ^{15}O production yields agree within 6.2 %. For both irradiated energies, the fitted ^{15}O yields are lower in the PMMA phantom than in the gelatine phantom.

In contrast to the gelatine phantoms, FLUKA simulations using external cross-sections slightly overestimate the production yields by 2.3 % to 5.3 % for the higher and the lower beam energy, respectively. This difference would in principle be tolerable, but since the trend of the discrepancy is in the opposite direction as for the gelatine phantom, an explanation for this effect is desirable. A difference due to neutron-induced nuclear reactions, that are not accounted for in the `fluscw.f` routine can be excluded. According to MC simulations, the neutron fluence is even higher in the PMMA phantom than in gelatine. It is therefore most likely, that this discrepancy occurs due to the general fit uncertainty and the lower production of ^{15}O in PMMA.

The highest radionuclide abundance in the PMMA phantom is given by ^{11}C. For this isotope the fitted yields are consistent with the findings from the other phantom materials. Due to the chemical composition of PMMA, where the fraction by weight of carbon is almost twice as high as the fraction by weight of oxygen, the largest amount of the total ^{11}C yield results from the (p,pn)-reaction from ^{12}C. Thus, the influence of uncertainties in the ^{16}O(p,3p3n)^{11}C reaction cross-sections is only a second order effect. This is supported by the only small underestimation of 2.3 % at most by MC simulations using experimental cross-sections.

The usage of the internal FLUKA models results in significant differences from the measurements. For ^{11}C and ^{15}O, under- and overestimations of up to 11.1 % and 37.2 % could be observed, respectively.

The third radionuclide, ^{13}N is not considered in the analysis of this phantom material due to its low production and the large fit uncertainties of 70 % and 128 % for 126.53 MeV and 178.01 MeV, respectively.

4.2.1.3 Comparison to previous data

Similar PET activation experiments have been done at other facilities. The yields obtained in the gelatine and PMMA phantoms at our facility are therefore compared to yields from HIT [Bauer] and GSI [Hildebrandt]. The production yields of both facilities are listed in table 4.4. It has to be stressed, that the experimental cross-sections used for FLUKA simulations in this work are the fine-tuned cross-sections from the experiments at HIT [Bauer et al., 2013]. However, differences between the published cross-section data and the cross-sections used in this work could be found for the ^{15}O production channel.

---

[4]According to Hildebrandt [2012], the fraction by weight of oxygen in PMMA is roughly 32 %, while in gelatine it is about 88.8 %.

HIT   Interestingly, the experimental ^{15}O yields from gelatine determined at HIT are about 7.6 % higher than the yields fitted at RPTC. They are thus almost 19 % higher than simulation results using these cross-sections (cf section 4.2.1.2). The major part of this discrepancy can be explained by the different cross-section data for this production channel. In contrast, the ^{11}C yields from gelatine are about 11.5 % lower as compared to the RPTC data, but they agree with simulation results within 2.3 %. The different chemical composition of their gelatine used in their experiments as compared to ours can only explain a few percent of this discrepancy (cf section 4.4.3). For ^{13}N, RPTC and HIT data agree within the large fit uncertainties. The largest difference between both fitted yields is 12.0 %.

For the PMMA phantoms, the comparison of the ^{15}O production measured in this work to data from HIT reveals similar findings as for the gelatine phantom, albeit the differences are a bit larger. Up to 16.6 % higher ^{15}O production yields were measured there. On the other hand, excellent agreement could be found for the ^{11}C production. Both data sets agree within 1.8 %.

GSI   The comparison to the in-beam data from GSI yields similar results as reported in Hildebrandt [2012] for the comparison of those data with offline data from HIT. Also in our experiments, an overall tendency of overestimation in the GSI analysis was observed. Indeed, except for the ^{11}C yield in the gelatine phantoms, where the difference is 12.4 % and 3.6 % for 126.53 MeV and 178.01 MeV, respectively, significantly higher production yields are extracted from in-beam PET measurements at GSI. A mean overestimation of 16.3 % for the gelatine phantom and even 42.6 % for the PMMA phantom could be found as compared to the offline measurements at RPTC. Hildebrandt [2012] explains these high discrepancies by uncertainties in the absolute calibration of the beam intensity and the detector efficiency.

4.2.2   Depth dependent absolute isotope yields

The knowledge of integral isotope yields is not yet sufficient for treatment plan verification by means of PET monitoring. Instead, the spatial distribution of the proton-induced β^+-activity is of major interest. Since a major issue in proton therapy is range uncertainty, a quantification of depth dependent isotope yields is necessary.

4.2.2.1   Comparison of the depth dependent isotope yields for the individual phantoms

Depth dependent isotope yields at the end of irradiation, extracted from the RPTC measurements, are plotted in figures 4.5, 4.6 and 4.7 for the three phantom materials PE, gelatine and PMMA, respectively. In order to compare the depth dependent isotope yields of this work to experimentally deduced yields at HIT, the simulation results using the

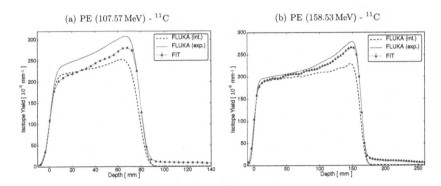

Figure 4.5: Depth dependent isotope yield of ^{11}C in the PE phantom: (a) corresponds to the irradiation with 107.57 MeV protons, (b) corresponds to 158.53 MeV. The dashed line and the solid lines without symbols correspond to the simulated isotope yields using FLUKA's internal model and the experimental cross-sections, respectively. The fit to the measured data is plotted by the triangles. The errorbars refer only to fit uncertainties, other uncertainties are not considered in this plot.

cross-sections validated and fine-tuned at HIT [Bauer et al., 2013] generally serve as a good substitution. Care has to be taken only for the [15]O yields, where the absolute yield determined at HIT and the FLUKA simulation using these cross-section data differ by up to 19 % (cf section 4.2.1.3).

PE   While the integral number of fitted ^{11}C-yield for the PE phantoms is in good agreement with FLUKA simulations using external cross-sections, the depth dependent isotope yield shows visible differences. The shapes of the depth profiles of measured and simulated isotope yields using experimental cross-sections are very similar. Sub-millimeter agreement of the distal fall-off can be seen for both energies. But the simulated profile exceeds the measured profile by nearly 10 % and 5 % for the lower and the higher beam energy, respectively.

On the other hand, a *tail* of isotope yield that ranges far beyond the penetration depth of the proton beam can be observed in the measured data, which is not present in the simulation results. For 107.57 MeV, about 5.2 % of the total fitted isotope yield can be found in this tail. The fraction is a bit smaller for the higher energy, but still clearly visible. The reasons for this unexpected isotope yields and its consequences will be discussed later in this section.

Gelatine   In the gelatine phantom, three different radionuclides are produced which obviously results in larger errorbars of the fit results. The profile of the ^{11}C production yield shows a pronounced peak in the beginning. This peak is the result of the 1 cm PMMA wall of the container into which the gelatine was filled. It is slightly lower than the peak predicted by both MC simulations. This might be due to the so-called *partial volume*

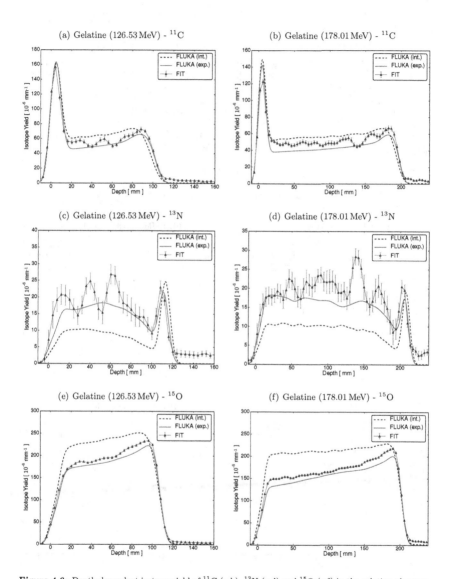

Figure 4.6: Depth dependent isotope yield of ^{11}C (a,b), ^{13}N (c,d) and ^{15}O (e,f) in the gelatine phantom. The left plots (a,c,e) correspond to the irradiation with 126.53 MeV protons, the right plots (b,d,f) to 178.01 MeV. The same color scheme as in figure 4.5 was used.

The strong fluctuation of the production yield in (c) and (d) is caused by low statistics and is further explained in the text.

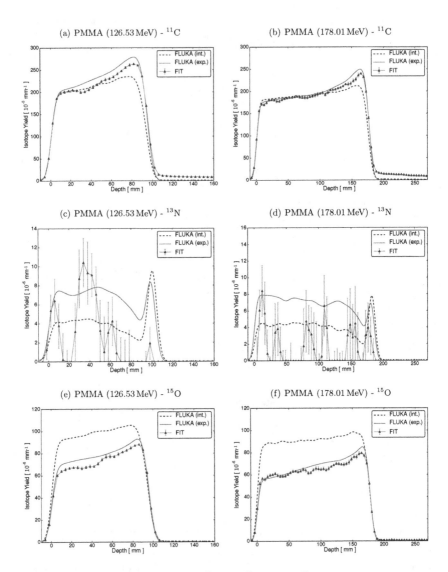

Figure 4.7: Depth dependent isotope yield of ¹¹C (a,b), ¹³N (c,d) and ¹⁵O (e,f) in the PMMA phantom. The left plots (a,c,e) correspond to the irradiation with 126.53 MeV protons, the right plots (b,d,f) to 178.01 MeV. The same color scheme as in figure 4.5 was used.

effect which occurs when a small, relatively high activity is measured next to a relatively low activity background. The effect will be explained when discussing the uncertainties in section 4.4. In gelatine, where $^{16}O(p,3p3n)^{11}C$ is the dominant production channel for ^{11}, the measured isotope yield is clearly higher than the yield calculated by FLUKA using experimental cross-sections. Concerning the distal fall-off, very good agreement could also be observed for this production channel.

The ^{13}N profile is strongly affected by bad statistics, since it is produced in the channel with the lowest production cross-section from the isotope ^{16}O. Moreover, because of its half-life which is in between the half-lives of ^{15}O and ^{11}C, the fitting routine has difficulties to accurately distinguish between $^{13}N/^{15}O$ and $^{13}N/^{11}C$. This can for instance clearly be seen at about 40 mm and 65 mm in the low-energy plots in figure 4.6, where the isotope yield of ^{13}N peaks and a corresponding valley can be observed in the ^{11}C profile. The characteristic peak of the ^{13}N yield at the end of the proton range which was predicted by simulations, could be reasonably well reproduced by measurements. This peak is already beyond the 50 % distal fall-off of the profiles of the other two isotopes and can therefore unambiguously be attributed to ^{13}N. However, slightly larger discrepancies of about 1 mm in the distal location of this peak, as compared to MC predictions, are measured here. Nevertheless, this is still tolerable.

For the depth profile of the dominant ^{15}O production yield, measurement agrees in shape with FLUKA predictions using external cross-sections. But for both beam energies and each depth, measured yields are higher than yields from the this MC simulation. This is what one would already expect from integral isotope yields calculated in the previous section.

The influence of the tail is a bit smaller than for the PE phantom but yet not negligible. According to PET measurements, about 4.1 % and 2.8 % of all produced isotopes are located in the tail for 126.53 MeV and 178.01 MeV, respectively.

PMMA All four production channels investigated in this study are combined in this phantom material. The results from the two pure phantoms can therefore be checked for consistency. For both beam energies, the ^{11}C profile of the PMMA phantom shows similar behavior as the PE profile. Yields calculated with FLUKA using experimental cross-sections and the measured profile agree in shape, but the predicted yields are significantly higher, while a relatively large fraction of ^{11}C is located in the tail, according to PET measurements. The small valley in the profile of the 126.53 MeV activation at about 40 mm again coincides with a large peak in the ^{13}N-profile and is therefore only a fit artifact.

For this phantom material, no conclusions can be drawn on the ^{13}N-production. Bad fit results due the relatively small fraction of produced ^{13}N lead only to a random occurrence of this positron emitter throughout the whole proton penetration depth. Even the characteristic peak at low proton energy could not be detected.

As already described in section 4.2.1, the calculated ^{15}O yield from FLUKA using

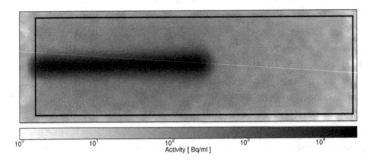

Figure 4.8: Logarithmic plot of the measured activity concentration of the gelatine phantom in the first time frame, summed over the perpendicular axis exiting from the page. The phantom boundaries are marked by the solid black lines. The PET scanner attributes a significant amount of activity to areas where activity is physically impossible.

experimental cross-sections are larger than the actually measured yields. Only the shape is similar, although the measurements might suggest a slightly smaller slope.

Similar to the gelatine phantom, the fraction of the produced isotopes in the tail is smaller than in the PE phantom. For the lower and higher energy, 4.1 % and 3.0 % of all radionuclides are located after the proton penetration depth, respectively.

4.2.2.2   FLUKA internal hadronic models

The well-known weakness of internal hadronic models of FLUKA 2011 can be seen in the presented data sets. The total amount of produced isotope yields differs by up to 37 % from the measured values. Also, the general shape of the depth-activity distribution, which can be seen in figures 4.5 - 4.7, is quite different from the experimental observations. These findings therefore agree with Seravalli et al. [2012] and Baumgartl [2014], who recommend to use experimental cross-sections instead of the internal models of MC codes, as already proposed by Parodi et al. [2002] for a former FLUKA version. However, the new FLUKA version which is announced to be released within the next few months should account for these discrepancies and lead to better results [Ferrari, 2014].

4.2.2.3   Experimental isotope yield beyond the proton penetration depth

One issue that has already been mentioned is the non-zero β^+-production yield found in the measured data at depths where one would actually not expect many nuclear reactions, since they are far beyond the penetration depth of the proton beam. A small fraction of this measured β^+-activity can be explained by nuclear reactions, induced by secondary neutrons. But it is unlikely that this contribution adds up such a large amount of activity (cf section 4.4.1). These tails, which can be seen in figures 4.5 - 4.7, carry up to 5.2 % of the total induced production yield. This is much more pronounced than the activity tails that can be seen in the depth-activity profiles in Bauer et al. [2013].

In figure 4.8, the measured activity concentration of the irradiated gelatine phantom is shown for the first time frame, summed over one axis. Activity concentration values are displayed using a logarithmic scale which helps to see even low activities. It can be seen, that the PET scanner attributes activity to voxels far beyond the penetration depth of the proton beam. But even in lateral positions outside the phantom, where there is only air, activity is displayed. Because of this, it can be excluded that the activity is only due to secondaries. It is rather likely an artifact introduced by the scatter correction of the PET scanner. However, replacing the scatter correction method *single scatter simulation*, which is according to the vendor the method of choice, by any of the other scatter correction methods results in even worse spatial distributions with more pronounced tails. It is therefore an additional source of uncertainty for the depth dependent isotope yields, which makes reliable quantification and therefore a fine-tuning of the experimental cross-sections difficult as it will be discussed in section 4.4.1.

4.2.3 Lateral activity profile

So far, only the depth activity profiles have been regarded. Nevertheless, also the lateral proton-induced activity distribution is of interest for PET monitoring. Information on the lateral spread of an irradiated field could be deduced from the measured transversal activity profile. This is especially useful for treatments where the proton beam paths are passing tangentially to organs at risk [Parodi, 2004].

Lateral widths of the β^+-activity profiles along the penetration depth are shown in figure 4.9 for three representative cases. For this analysis, two-dimensional Gaussians were fitted to individual depth-slices of activity distributions measured in the first time frame of PET acquisition, in order to determine both widths $FWHM_x$ and $FWHM_y$. The same was done for data sets derived from FLUKA simulations via eq. (4.7), corresponding to the activity of the same time frame.

Although no severe discrepancies between widths determined from MC simulations and from PET measurements could be seen, no clear conclusion on the overall agreement could be drawn. Basically, three different types of results were obtained (cf figure 4.9), none of them occurring more often than the other types. Also, no dependence on the phantom material, the beam energy or the direction in the PET scanner could be found.

Figure 4.9 (a) shows one example where excellent agreement between the expected and measured lateral dimensions of the activity distribution could be observed. In figure 4.9 (b), small differences in the width of the lateral activity profile can be seen. However, measurement and simulation show similar broadening with penetration depth. Figure 4.9 (c) shows a completely different profile. The FWHM of the measured activity profile is convex with the broadest width somewhere in the plateau, whereas it slightly decreases towards the location of the Bragg peak. This kind of profile resembles a lot the depth-profile that

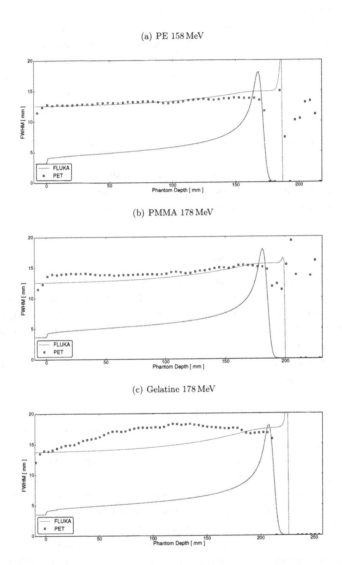

Figure 4.9: Exemplary plots of the FWHM of the lateral β^+-activity profiles along the penetration depth. The solid lines are the widths calculated by FLUKA, the squares refer to the fits on the measured activity. To guide the eye, the dotted black line is the depth-dose profile in arbitrary units.

can be seen in figure 3.5 (c). For that plot, a double Gaussian was fitted to the lateral dose profile of a single pencil beam at various depths. The broad Gaussian there is assumed to correspond mainly to secondary particles produced inside the phantom in inelastic collisions. This convex shape therefore indicates, that the single Gaussian approximation, which was used here, is not the best method to describe the lateral activity profile. A better description could probably be achieved by fitting a two-dimensional double Gaussian distribution to the lateral profiles. However, deeper analyses would be beyond the purpose of this work.

Finally, it has to be stressed that the agreement between simulation and measurement in the upper plot of figure 4.9 strongly depends on the choice of the FWHM for the Gaussian filter that was used to model the finite imaging resolution of the PET scanner (cf section 4.1.4.2). This was assumed to be constant all across the FOV of the PET scanner, which is only an approximation. More reliable results could be achieved by a space-dependent FWHM for the Gaussian filter, since the imaging resolution of the PET scanner is better in the center of the FOV than in the outer regions.

4.3   Additional investigation of the PET scanner performance

Within the time dependent analysis for the quantification of the isotope yields (cf section 4.2.1), it could be observed that the fitted activity for a late time frame is lower than the measured activity, while the opposite is found for the first time frames with high activity (cf figure 4.4). It therefore appeared that there is no stringent linearity of the measured activity with the true activity, but rather that the PET scanner underestimates the activity at high activity concentrations. This assumption deserves to be investigated in more detail to assess whether the scanner performance could have a non-negligible influence on the reconstructed activities and the resulting production yield estimates.

4.3.1   Test experiments with FDG

In order to get a first insight on the PET scanner performance depending on the activity level, a simple experimental setup that allows comparing the activity measured by the PET scanner to the real activity, measured with an independent method, is desired. Of course, calibration measurements concerning the absolute activity of the PET scanner are performed on a regular basis, but the active volume used for these measurements is significantly larger than the active volume in our activity measurements. It is therefore best to design an experimental setup that has a similar activity concentration and distribution as for the activation experiments

Thus, a small bottle was filled with 100 ml NaCl solution and fluorodeoxyglucose (^{18}F-FDG, or FDG), which has a half-life of $T_{1/2}(\text{FDG}) = 109.77$ min. Before injecting the FDG into the bottle, the activity of the syringe containing the FDG was measured with an activimeter, calibrated for this kind of syringes. In order to account for the residual

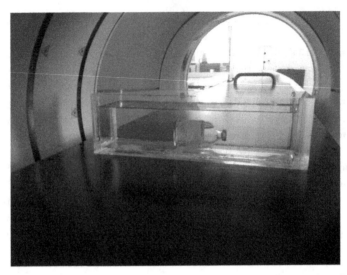

Figure 4.10: Setup of the FDG test measurement to evaluate the scanner performance depending on the activity level. A small bottle filled with a known activity is placed in the water-filled PMMA container. The PET scanner acquires the activity during 6 h measurement time.

activity which is left in the injection after adding the FDG into the bottle, its activity was determined in the activimeter as well. The small bottle was then sticked to the bottom of the PMMA container that was also used for the gelatine activation experiments. The container was then filled with approximately 2.5 l of water in order to exclude differences due to the different attenuation and scatter corrections. In figure 4.10 the setup of this experiment is shown.

The true activity $A_{\text{true}}(t)$ can be determined accurately for any time by considering the time delay between the two activimeter measurements Δt and the delay between the second activimeter measurement and the time of the PET acquisition t:

$$A_{\text{true}}(t) = \left(A_{\text{all}} \cdot \exp\left(-\frac{\ln 2}{T_{1/2}(\text{FDG})}\Delta t \right) - A_{\text{residual}} \right) \exp\left(-\frac{\ln 2}{T_{1/2}}t \right). \qquad (4.8)$$

A_{all} and A_{residual} are the activities of the full and the empty syringe, respectively, measured with the activimeter.

For the PET acquisition and reconstructions, the same parameters as described in section 4.1.2 were used, except the acquisition time was set to 6 h, which allows comparison of the activity over more than 3 half-lives of ^{18}F. The comparison between expected activity and activity measured with the PET scanner was done at three different delay times. To this aim, three 1 min long time frames were reconstructed (cf table 4.5). The total measured activity for each of these time frames was obtained by integrating the reconstructed activity distribution over the complete scanned volume, in contrast to the data analysis of the PET

Table 4.5: True activity and integrated activity measured by the PET scanner for the FDG-sample at 3 different time points. The true activity was measured with a calibrated activimeter at $t_{del} = 0$. The difference between true and measured activity is shown in the last column.

t_{del} [min]	$A_{true}(t_{del})$ [MBq]	$A_{PET}(t_{del})$ [MBq]	ΔA
13.5	18.50	15.23	17.7 %
234.5	4.58	4.09	10.7 %
372.5	1.92	1.78	7.3 %

activation experiments, where only the phantom volume with a small margin around was used for the evaluation (cf section 4.1.4.1).

In table 4.5 the activities and their relative differences are summarized. Although only meant to give a first impression on the scanner performance, important results could be obtained. The results indicate, that the scanner significantly underestimates the activity for small volumes at high activity concentration. The difference decreases with decreasing activity. The high activity discrepancy in the first evaluated time frame is, however, not of concern for the activation measurements, because the activity concentrations determined in the phantom experiments are always lower than in the beginning of the FDG-measurement. Nevertheless, the activities in the second and third time frame reflect typical activities that could be achieved in the first time frames of the activation experiments, especially for the gelatine and the PMMA phantoms where the contribution of the short-lived ^{15}O yields high activities within the first few minutes of data acquisition.

Concerning the accuracy of this estimation, three sources of uncertainty have to be evaluated. First of all, the time delay between activation measurement and PET acquisition start was determined by the time stamp function of the activimeter, which is located in the room next to the PET/CT scanner. Therefore, the uncertainty in the delay time is assumed to be 0.5 min. Moreover, the intrinsic uncertainty of the activimeter measurement of the full and the empty syringe, as well as the error that can occur due to the injection of FDG to the small bottle have to be taken into account. But even for a conservative estimate of 5 % for the uncertainty of the initial true activity and taking into account the uncertainty of the time delay, A_{true} would still be 5.8 % and 2.3 % larger than A_{PET} for the second and the third time frame, respectively. A further source of uncertainty is due to the attenuation and scattering correction for the bottle. This systematic error is constant for all three evaluation time frames and might be large enough to explain the remaining disagreement, at least in the last evaluated time frame.

Unfortunately, these results cannot directly be transferred to the activation experiments. The activity distributions of both experiments are similar, yet there are still some important differences. While the active volume of both experiments is comparable, the length of the bottle is smaller and the diameter is larger than the typical distal and lateral profile of the irradiation-induced activity. Also, the activity distribution inside the bottle is homogeneous, in contrast to the distribution of the induced β^+-activity.

Therefore, in two further experimental sessions, we irradiated the PE and the gelatine phantoms with different numbers of protons. In the following two sections, the results of the experiments are presented.

4.3.2 PE

Due to the half-life of about 20 min, the activity contribution within the complete PET acquisition by ^{11}C does not change as much as the contribution by ^{15}O. In fact, after 36 min, which was the duration of the PET acquisition for this phantom material, the activity is still about 30 % of its initial activity. This makes the PE phantom, where ^{11}C is the only detectable radionuclide produced by proton irradiation, a good candidate for an evaluation of the influence of the dose level on the fitted isotope yield.

In a further experimental session, the PE phantoms were irradiated with a 158.53 MeV proton beam with twice and three times the proton number as compared to the initial activation experiment (cf table 4.2). For the sake of clarity, the phantom irradiation with the lowest amount of protons will be hereafter called PEx1, the irradiation with twice as much protons PEx2 and the irradiation with the highest amount of protons PEx3.

Moreover, several new fit results were obtained for all 3 measurements by varying the time frames selected for the fitting routine. In this way, irradiation with lower doses could be simulated from existing data sets. Of course, the time delay between irradiation and the first frame included for the fit cannot be increased arbitrarily. At an additional time delay of about 22 min, the statistics starts getting noticeably worse, resulting in increased fit uncertainties. Also, this method can not be applied to the data sets of the other phantoms because of the different half-lives of the different β^+-emitters involved there. The time frames for the reconstructions were 36×1 min for PEx1 and 24×1.5 min for PEx2 and PEx3.

Before evaluating the results, the data sets from the three independent experiments had to be checked to be consistent. Therefore the first 21 min of the PEx2 data set were skipped. The remaining 15 min of this data can now be seen as a data set of a fictive experiment, irradiated with about the same amount of particles as the PEx1 data set. This has been also done similarly, but skipping only 12 min, for the PEx3 data set to be comparable to the PEx2 data set. It turns out, that PEx1 and PEx2 agree within 2.7 % and PEx2 and PEx3 agree within 0.8 %. Part of the difference between PEx1 and PEx2 can be attributed to the different selection of time frames (cf section 4.2). Thus, the reproducibility of the activation experiments could be shown.

The fitted ^{11}C-production yields for the different irradiation parameters and different delay times indicate, that there is a non-negligible dependence on the true activity. Since the scanner tends to underestimate the activity at high activity concentrations (cf section 4.3.1), the estimated yields become smaller when more particles are delivered to the phantom. This can be seen in the depth-dependent profile of the isotope yield in figure 4.11.

In table 4.6, calculated yields for some frame selections are tabulated. When the

Table 4.6: Influence of the true activity level on the scanner performance. The 3 data sets correspond to 3 activation experiments under the same conditions but different numbers of delivered protons. The second column is the fitted total isotope production yield, including the whole acquisition time. The third and the forth column are the fitted yields including only the first half and the second half of the measured data set in the fitting procedure, respectively. In the last column, the differences between the the fitted yields including the first and the last half of the data set is given. In the last row, the relative differences between the fitted yields of PEx1 and PEx3 are shown. In order to eliminate the influence of the different durations of the irradiations, the isotope yields given in this table are yields assuming instantaneous production.

Data set	Total production yield per 10^6 protons			Difference 1st/2nd half
	all frames	1st half	2nd half	
PEx1	37479	37264	37924	1.8 %
PEx2	35860	35591	36424	2.3 %
PEx3	35233	34965	35799	2.4 %
Difference PEx1/PEx2	6.0 %	6.2 %	5.6 %	

measured data from the complete PET acquisition time (36 min) is included in the fit, the production yield of the PEx1 experiment is 6.0 % larger than for the PEx3 experiment. This is significantly larger than the differences in the consistency check between PEx1, PEx2 and PEx3 as described above. But even within each experiment, the fitted yields differ by up to 2.4 %, depending whether the first 18 min or the last 18 min of the total acquisition are included in the analysis. Thus, the data analysis of the first 18 min of PEx3, corresponding to a total delivery of 3.15×10^{11} protons, results in a ^{11}C production yield that is 7.8 % smaller than the yield obtained from the analysis of the last 18 min of PEx1, which corresponds to a delivery of 5.69×10^{10} protons. The differences can also be seen in the semi-logarithmic plot in figure 4.12.

These results carry an important message for the desired validation of cross-sections against measurement for the implementation of PET monitoring. Generally, one is tempted to keep the number of delivered protons as high as possible in order to get good statistics and hence nice fit results. However, since the activated volume is rather small as compared to the volumes usually investigated by means of PET, and the activity concentration is generally high in the irradiated volume, dead time losses of the detectors can result in an underestimation of the absolute isotope production yields, depending on the effectiveness of the scanner dead time correction. In our specific activation experiments, differences of 7.8 % and possibly even more can be attributed to an improper choice of irradiation parameters. It is therefore necessary to find an adequate a trade-off between increased fit uncertainties due to low statistics and an increased error due to the scanner dead time losses.

4.3.3 Gelatine

The trade-off mentioned in section 4.3.2 could be of special relevance for the choice of the irradiation parameters for the gelatine activation. Due to the high fraction of produced ^{15}O in this phantom, measured activity varies within one order of magnitude within one PET

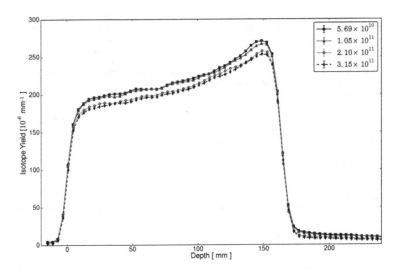

Figure 4.11: Depth-dependent isotope yield for different numbers of delivered protons in PE. The dark gray, light gray and the dashed black lines correspond to the irradiations PEx1, PEx2 and PEx3, respectively. For the solid black line, only the second half of the time frames of PEx1 was included in the fitting procedure. The proton numbers in the legend are the delivered protons, calculated as described in section 3.3, except for the solid black line. In this case, the proton number of PEx1 was lowered taking the additional decay of 18 min into account. Again, the influence of the different durations of the irradiations was eliminated by assuming instantaneous production.

acquisition. Especially the fitted production yield of ^{15}O is expected to be underestimated because the first few time frames with high activity, which are decisive for the results, are affected most by dead time effects.

However, as reported in section 4.2, the fit uncertainties for the gelatine phantom are already intrinsically large due to counting statistics. Hence, lowering the number of delivered protons by a too high grade would eventually make the fit results useless. From the previous analysis for PE and the measured activities in the first time frames of the former gelatine measurement, the best would be to reduce the delivered dose to about 1/7-th. However, assuming that the uncertainty scales with $\sqrt{N}/N$, this would increase the fit uncertainties to over 10 % for the ^{11}C production yield. As a reasonable compromise, the number of delivered protons was therefore lowered to about 30 % of the initial value.

According to the findings in section 4.3.2, one would expect the largest differences for the ^{15}O production. This could be indeed confirmed by measurements. Using only 30 % of the protons initially delivered in the activation experiments results in about 3.2 % and 0.8 % higher ^{15}O production yields for the 126.53 MeV and 178.01 MeV irradiation, respectively. On the other hand, the difference in the fitted production yields of ^{11}C and ^{13}N are within the intrinsic fit uncertainties. The ^{11}C production yield might be lower, but actually no clear trend could be observed for these two isotopes. All fitted isotope yields

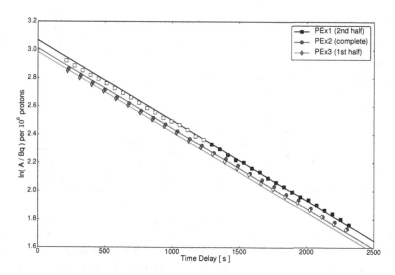

Figure 4.12: Semilogarithmic plot of the activity versus time delay from the end of irradiation. Squares, circles and diamonds correspond to the total measured activities in each time frame for the PEx1, PEx2 and PEx3 irradiation, respectively. The lines are the fitted activities for these three experiments. Only filled symbols were included in the data set for the fit. The empty symbols are data points not included in the fit. Thus, for the dark and the bright line, only the last and the first 18 min of the data acquisition were included in the fit, respectively.

and the difference between the yields of both irradiations can be found in table 4.7.

Concerning the low-activity tail described in section 4.2.2, only a minor increase of its fractional contribution to the total activity could be detected as compared to the irradiation with the higher number of protons. At 126.53 MeV, the fraction of fitted activity in this tail is raised from 3.9 % to 4.8 % of the total activity, whereas for the 178.01 MeV irradiation, the ratio stays almost constant (minor raise from 2.5 % to 2.7 %).

In conclusion, dead time effects of the scanner have a small, but non-negligible influence on the fitted production yields of ^{15}O. Up to 3.2 % higher yields could be detected by a reduction of the delivered proton number. Hence, it is desirable to keep the number of delivered protons low enough to make dead time losses of the scanner negligible. This, of course, significantly deteriorates the statistics and increases the fit uncertainty for the ^{11}C and ^{13}N production yields. It might therefore be useful to determine the cross-sections of the ^{16}O(p,3p3n)^{11}C and the ^{16}O(p,2p2n)^{13}N reactions separately by irradiating the phantom with a higher number of protons and prolong the time delay between irradiation and measurement to rule out the ^{15}O contribution.

However, when offline PET monitoring is used in clinical practice, the time delay between irradiation and PET acquisition start would be increased as compared to the phantom activation experiments. For the transport and especially repositioning of the patients, 10 min delay, as assumed in the late-offline PET concept in Hildebrandt [2012], seems much

Table 4.7: Fitted isotope production yields and their relative fit uncertainties for the gelatine phantoms at both energies. *High activity* and *low activity* refer to the irradiations using 100 % and ≈ 30 % of the initial amount of protons (cf section 4.2.2), respectively. The difference between the yields of both experiments is shown in the last column. The influence of the different durations of the irradiations was eliminated by assuming instantaneous production.

Energy [MeV]	Isotope	high activity	low activity	Difference high/low
126.53	^{11}C	7739 ± 5.52 %	7538 ± 8.36 %	−2.6 %
	^{13}N	2434 ± 18.74 %	2493 ± 25.98 %	6.6 %
	^{15}O	22060 ± 1.45 %	22756 ± 1.91 %	3.2 %
178.01	^{11}C	11649 ± 5.24 %	11675 ± 8.42 %	0.2 %
	^{13}N	4393 ± 14.57 %	4057 ± 26.47 %	−7.7 %
	^{15}O	35670 ± 1.12 %	35957 ± 2.12 %	0.8 %

more realistic than ≈ 3 min needed for the phantom experiments. It is therefore likely, that the influence of an underestimation of the ^{16}O(p,pn)^{15}O reaction cross-section would not have any noticeable impact on the plan verification in offline PET monitoring. Yet, for in-room PET monitoring where the delay is much shorter, the influence of the scanner dead time should be taken into account when validating experimental cross-sections.

Moreover it has to be stressed, that in clinical applications the activity concentration will be much lower. The delivered dose in one treatment fraction is typically about one order of magnitude smaller and the irradiated volume is larger than in the presented PET activation experiments. Furthermore, biological washout will additionally lower the activity concentration.

4.4   Discussion of uncertainties

Before going to the conclusions, a brief analysis of the major uncertainties that belong to the quantification of the fitted isotope yields is necessary. These several sources of uncertainties can be classified in three groups: uncertainties due to the PET scanner, the data analysis and further uncertainties.

4.4.1   Scanner performance

The largest uncertainties in the quantification of positron emitter yields are due to the PET scanner itself. Apart from the relatively low spatial resolution which is an intrinsic problem of PET imaging and is therefore not further discussed, the following sources of uncertainties contribute to the total uncertainty:

Alignment of the phantom in the scanner. Both in axial as well as in trans-axial direction, the spatial resolution of a PET scanner decreases with increasing distance to the center of the FOV. One can therefore not exclude, that the alignment of the activated phantom in the PET scanner has an influence on the measurement. Unfortunately, it was not possible throughout this work to test the influence of the phantom orientation inside the FOV (axial / trans-axial). We therefore rely on the findings of Bauer et al. [2013], where no orientation dependence could be observed at their PET scanner (Siemens Biograph mCT).

Partial volume effect. Due to the limited spatial resolution of PET scanners, small *hot* spots relative to *cold* background show partial loss of intensity, while their activity is smeared out over a larger area [Saha, G.B., 2010]. This effect is commonly referred to as partial volume effect. This effect is only relevant for the gelatine phantom, where the beam passes through 1 cm PMMA before it enters the gelatine. In late time frames, the activity of the gelatine is already much lower than the activity in the PMMA. The scanner therefore underestimates the activity of the PMMA which leads to a lower predicted ^{11}C production yield in the PMMA entrance window (cf section 4.2.2).

Scatter correction. It has already been mentioned in section 4.2.2 that there are some artifacts introduced in the PET image due to suboptimal performances of the scatter correction method applied to the data. Up to 5.2 % of the fitted β^+-emitter yield is beyond the actual penetration depth of the protons and a *glowing* in the image can be observed at locations outside the phantom. A small fraction of the spatially uncorrelated activity background can be explained by nuclear reactions induced by secondary neutrons. According to FLUKA simulations using the internal hadronic model, up to 0.5 % of the total induced activity can be found in a tail beyond the penetration depth of the protons and can thus be ascribed to secondary neutrons. However, neutron-induced β^+-activity cannot explain activity outside the phantom. But since these events were actually measured and only attributed to wrong voxels, one cannot simply subtract them from the image as it would lead to an underestimation of the production yields. This 5.2 % is therefore a spatially uncorrelated amount of positron emitter yields which has to be considered as uncertainty for the depth dependent isotope yield.

CT attenuation correction. The CT attenuation correction for the annihilation phantoms is of course optimized for human tissue. But this bilinear calibration curve for the conversion of CT number to attenuation coefficients for 511 keV photons assigns higher attenuation coefficients to PE and PMMA, as compared to literature data from XCOM photon cross-section database [Berger et al., 2010]. The attenuation correction method therefore assumes, that annihilation photons in PE and PMMA are 4.1 % and 6.8 % more attenuated, respectively. This then yields to a slight overestimation of the activity in PE and PMMA. For gelatine, the attenuation coefficient of the PET scanner's calibration curve is in good agreement with data from XCOM database.

Dead time losses. This effect has been studied in detail in section 4.3 since it is probably the major source of uncertainty of these activation studies. Up to 3.2 % and 7.8 % difference could be observed for the gelatine and the PE phantom, respectively. It is therefore necessary to make a trade-off between high activity in order to get good fit results and low activity in order to keep the effect of detector dead time losses small.

4.4.2 Fit uncertainties

There are basically two sources of uncertainties arising due to the data analysis. Since ^{22}Na was selected for decay correction, the activities stored in the pixel values of the PET DICOM files are the average activities of the time frame and are set to the midpoint of the frame duration. But because of the exponential decay of the radionuclides, the time of the average activity of a time frame, t_{av}, is always sooner than the midpoint of the frame duration t_{frame}:

$$t_{av} = \frac{1}{\lambda} \ln \left(\frac{\lambda t_{frame}}{1 - e^{-\lambda t_{frame}}} \right) . \tag{4.9}$$

One could account for this when there is only one radionuclide present, but if there is more than just one isotope, the ratio of the initial activities of each isotope has to be known in order to solve the equation. From eq. (4.9) one can estimate the uncertainty due to the usage of the midpoint of each time frame for the fit. The error introduced is negligible for the PE phantom, but it becomes larger for the gelatine and the PMMA phantoms, especially during the first minutes of PET acquisition where ^{15}O is the dominant β^+-emitter. For a frame duration of 1 min, the time of the average activity in a time frame is almost 1 s before the midpoint.

The other fit-related source of uncertainty is due to the selection of time frames, which had already been discussed in section 4.2.1.1. Disregarding variable time frame lengths because of their negative influence on the convergence of the fit [Bauer et al., 2013], the largest difference in fitted yields due to the selection of time frame length is 1.1 % in this work.

4.4.3 Further uncertainties

There is still a large amount of further sources of uncertainties, although most of them are rather small as compared to the major uncertainties mentioned before.

MU calibration. The exact number of protons delivered to the phantom is not known but has to be calculated from the prescribed Monitor Units. In section 3.3, the conversion formalism was presented and its uncertainty was estimated to be 1.5 %.

Cutting out phantom. Only the part of the PET image where the phantom was located and a small margin around was taken for data evaluation. However, due to the suboptimal scattering correction, typically around 5 % of the total activity that was measured by the PET scanner was attributed to voxels outside the phantom and the small margin. This results in a spatially uncorrelated activity background that extends over the complete PET acquisition volume.

Gelatine composition. Instead of water, gelatine was used for the analysis of the three proton induced reaction channels with ^{16}O as mother isotope. Introducing this agar-agar powder into the water opens the additional isotope production channel ^{12}C(p,pn)^{11}C. Sommerer et al. [2009] judged the impact to be negligible for in-beam PET measurements. However, for offline measurements, where the fraction of activity due to the short-lived ^{15}O is much smaller, the impact might be larger. Therefore, in the material definition of the simulation input, the chemical composition of gelatine is set to $H_{66.2}O_{33.1}C_{0.7}$ as described in Sommerer et al. [2009] and recalculated for the mixture used in the experiments. According to FLUKA simulations, about 8.9 % of the total ^{11}C yield in gelatine is produced in ^{12}C(p,pn)^{11}C-reactions. Increasing the carbon content in the material definition by 10 % changes the ^{11}C production yield by less than 1 %. The uncertainties in the chemical composition of gelatine can thus be judged to be negligible as compared to other sources of uncertainty.

Variation in beam delivery. In the data analysis, constant beam delivery is assumed during activation. In reality, however, fluctuations in the beam intensity are unavoidable. But since the duration of the irradiation is much shorter than the half-lives of any considered isotope, uncertainties due to these fluctuation can be neglected.

Time measurement. Also the measurement of beam-on time and delay between irradiation and measurement is a potential source of uncertainty. The method used in the experiments, involving the video function of a camera and an offline video analysis for extracting the times enables sub-second precision. An error in time measurement in the order of several tenths of seconds is definitely negligible as compared to the uncertainties of the other two groups.

For comparison, Bauer et al. [2013] evaluated the impact of a variation of ± 5 s in the time delay, which is much higher than the variations to be expected in the present work, and finds a difference of 1.4 % in isotope yield.

5 Conclusion and Outlook

In this work, two crucial steps towards an implementation of offline PET monitoring at the Rinecker Proton Therapy Center were done. A Monte Carlo model of the clinical proton beamline as described in Englbrecht [2014] was further developed regarding the correct modeling of the proton beam spot scanning and the absolute dose output. The influence of the field size on the dose output was investigated for measurements, MC simulations and the predictions of a conventional, analytical dose calculation algorithm. Furthermore, activation experiments for the validation of experimental cross-sections for the dominant positron emitter production reactions were performed. The proton-induced β^+-activity of three different phantom materials with similar chemical composition as human tissue was measured with a conventional full-ring PET/CT scanner, which is located close to the treatment rooms. In this context also the performance of the PET scanner at high activity concentrations was evaluated.

The first ingredient for PET monitoring could be successfully achieved. The MC model of the clinical beamline works fine and results fulfill clinical requirements concerning the agreement in range, spot size [Englbrecht, 2014] and absolute dose. The accuracy of the absolute dose calibration can be specified to be within 1.5 %, which is comparable to the accuracy of about 1.0 % with respect to standard geometry and dosimetric instrumentation in the clinic. It is moreover possible to simulate arbitrary treatment plans due to the implementation of gantry and table motions as described in Kopp [2014]. Valuable insights could be found concerning the influence of the low-dose envelope of scanned proton beams. Especially for small fields, significant discrepancies between the dose calculated with the planning software and the delivered dose are possible. In contrast, dose calculated with FLUKA was in good agreement with measurements. In critical cases involving small fields it is therefore favorable to verify the predicted dose by measurements or MC simulations. However, the findings of this work also showed that these differences generally tend into directions that have no negative influence on the treatment. The real dose in healthy tissue is thus lower than the predicted one whereas no significant differences could be detected in the SOBP i.e., the tumor region.

The validation of experimental cross-sections could not be completed successfully. Although the integral proton-induced production yields measured at the RPTC are similar to findings at HIT, with mean differences of 8.5 % for gelatine and 6.8 % for PMMA, no clear results concerning the depth dependent yields could be found due to encountered technical limitations of the PET/CT scanner used in this study. In fact, up to 5.2 % of the total measured activity was located beyond the penetration depth of the proton beam. Even at positions outside the phantom, activity was measured by the PET scanner. This is most likely due to problems of the intrinsic scatter correction methods. For this reason, fine-tuning of the production cross-sections would be meaningless.

Anyway, an important finding could be obtained within this analysis. Care has to be taken when performing the cross-section validation due to dead time losses of the PET scanner. For activation experiments one is tempted to deliver a large number of protons to the phantom in order to get good count-rates, leading to good fit results. It could however be shown, that despite a relatively low integral induced activity, this results in activity concentrations in the beam path which are higher than typical concentrations in diagnostic cases. In that activity range, the dead time correction applied by the scanner during reconstruction apparently does not work properly anymore. High activities are therefore underestimated, which would lead to underestimations in the production yields and consequently in the cross-sections. Up to 7.8 % differences in the production yield could be observed within this work. Since this effect is not constant over the acquisition time and the volume, it cannot be corrected for by any global scaling. A careful trade-off between high count-rate and dead time losses has therefore to be made when validating cross-sections for PET monitoring. However, newer PET scanners based on LSO or LYSO crystals could possibly show a less pronounced dependency on the strength of the activity concentrations due to the shorter dead time of these scintillators.

In conclusion, the PET/CT scanner which was used in this work does not serve as an optimal basis for offline PET monitoring. The issue of measured activity at locations where induced activity is physically impossible must be solved before using this method in clinical routine. But if one is only interested in the proton range and the shape of the activity profile, while no quantitative information on the activity is needed, promising results can be obtained in prospective studies. Instead of homogeneous phantoms as used in this work, proton range and lateral profiles in heterogeneous or even anthropomorphic phantoms can be studied in future investigations. The problem of dead time would probably not occur at lower activity levels, comparable to the resulting activity levels in patient treatment.

For quantitative studies or for implementing offline PET monitoring in clinical routine, enhanced technical specifications of a newer PET/CT scanner would be advantageous. Phantom studies with the same homogeneous phantoms are then again necessary, but considerably less time will be needed for this because of the know-how gained within this present work. Like this, also different phantom materials for other β^+-production reactions have to be studied. After the successful validation and fine-tuning of all relevant production cross-sections, another issue that has not been mentioned yet has to be solved - the biologic washout when irradiating living tissue [Parodi et al., 2007, Bauer et al., 2013,

Ammar et al., 2014]. But with all this solved, it may be possible to further improve proton therapy and possibly decrease the size of the safety margins for the full exploitation of the clinical benefits of proton therapy.

Bibliography

NuDat 2 databaseNational Nuclear Data Center. Available: http://www.nndc.bnl.gov/nudat2/ [2014, October 7].

RPTC. Available: http://www.rptc.de [2014, October 7].

C. Ammar, K. Frey, J. Bauer, C. Melzig, S. Chiblak, M. Hildebrandt, D. Unholtz, C. Kurz, S. Brons, J. Debus, A. Abdollahi, and K. Parodi. Comparing the biological washout of β+-activity induced in mice brain after 12C-ion and proton irradiation. *Phys. Med. Biol.*, 59(23):7229–7244, Dec 2014.

F. H. Attix. *Introduction to Radiological Physics and Radiation Dosimetry*. A Wiley-Interscience publication. Wiley, 1986. ISBN 9780471011460.

C. Bai, L. Shao, A. Da Silva, and Z. Zhao. A generalized model for the conversion from CT numbers to linear attenuation coefficients. *Nuclear Science, IEEE Transactions on*, 50(5):1510–1515, Oct 2003. ISSN 0018-9499.

W. Barkas and D. Evans. *Nuclear Research Emulsions: Techniques and theory*. Pure and applied physics. Academic Press, 1963.

G. Battistoni, S. Muraro, P. Sala, F. Cerutti, A. Ferrari, S. Roesler, A. Fassó, and J. Ranft. The FLUKA code: Description and benchmarking. In M. Albrow and R. Raja, editors, *Proceedings of the Hadronic Shower Simulation Workshop 2006*, pages 31–49. Fermilab, Sept. 2006, 2007.

J. Bauer. Experimental positron emitter yields at HIT. private communication.

J. Bauer, D. Unholtz, C. Kurz, and K. Parodi. An experimental approach to improve the Monte Carlo modelling of offline PET/CT-imaging of positron emitters induced by scanned proton beams. *Phys. Med. Biol.*, 58(15):5193–213, Aug. 2013.

J. Bauer, D. Unholtz, F. Sommerer, C. Kurz, T. Haberer, K. Herfarth, T. Welzel, S. E. Combs, J. Debus, and K. Parodi. Implementation and initial clinical experience of offline PET/CT-based verification of scanned carbon ion treatment. *Radiother. Oncol.*, 107(2): 218–226, MAY 2013. ISSN 0167-8140.

M. Baumgartl. PET-based Hadrontherapy Monitoring: Monte Carlo Simulations for Nuclear Interaction Studies in Phantoms. Master's thesis, LMU München & Université Paris Sud, 2014.

J. Beebe-Wang, P. Vaska, F. A. Dilmanian, S. Peggs, and D. Schlyer. Simulation of proton therapy treatment verification via PET imaging of induced positron-emitters. In *Nuclear Science Symposium Conference Record, 2003 IEEE*, volume 4, pages 2496–2500 Vol.4, Oct 2003.

M. Berger, J. Coursey, M. Zucker, and J. Chang. ESTAR, PSTAR, and ASTAR: Computer Programs for Calculating Stopping-Power and Range Tables for Electrons, Protons, and Helium Ions (version 1.2.3), 2005. Available: http://physics.nist.gov/Star [2014, September 20].

M. Berger, J. Hubbel, S. Seltzer, J. Chang, J. Coursey, R. Sukumar, D. Zucker, and K. Olsen. XCOM: Photon Cross Section Database (version 1.5), 2010. Available: http://physics.nist.gov/xcom [2014, October 7].

H. Bethe. Zur Theorie des Durchgangs schneller Korpuskularstrahlen durch Materie. *Annalen der Physik*, 397:325–400, 1930.

M. Durante and J. S. Loeffler. Charged particles in radiation oncology. *Nat. Rev. Clin. Oncol.*, 7(1):37–43, JAN 2010. ISSN 1759-4774.

XiO Proton Pencil Beam Scanning System - Beam Data Requirements for Beam Modeling. Elekta, Stockholm, Sweden, 2012.

W. Enghardt, P. Crespo, F. Fiedler, R. Hinz, K. Parodi, J. Pawelke, and F. Ponisch. Charged hadron tumour therapy monitoring by means of PET. *Nucl. Instr. Meth. Phys. Res. A*, 525(1-2):284–288, JUN 1 2004. ISSN 0168-9002. International Conference on Imaging Techniques in Subatomic Physics, Astrophysics, Medicine, Biology and Industry, Stockholm, SWEDEN, JUN 24-27, 2003.

F. Englbrecht. Monte Carlo modeling to support quality assurance at a clinical proton therapy facility. Master's thesis, Ludwig Maximilians Universität München, 2014.

A. Ferrari. FLUKA: from LHC to medical applications, June 2014.

A. Ferrari, P. R. Sala, A. Fassó, and J. Ranft. FLUKA: a multi-particle transport code. *CERN-2005-10*, INFN/TC_05/11, SLAC-R-773, 2005.

F. Fiedler, D. Kunath, M. Priegnitz, and W. Enghardt. Online Irradiation Control by Means of PET. In U. Linz, editor, *Ion Beam Therapy: Fundamentals, Technology, Clinical Applications*, chapter 31. Springer, 2011. ISBN 9783642214141.

Advanced Geometry, FLUKA beginner's course lecture notes, 2013. FLUKA. 14th FLUKA Course, Dresden, Germany.

L. Grevillot, D. Bertrand, F. Dessy, N. Freud, and D. Sarrut. A Monte Carlo pencil beam scanning model for proton treatment plan simulation using GATE/GEANT4. *Phys. Med. Biol.*, 56(16):5203–19, Aug. 2011.

V. Highland. Some practical remarks on multiple-scattering. *Nuclear Instruments & Methods*, 129(2):497–499, 1975. ISSN 0029-554X.

M. Hildebrandt. Experimental positron emitter yields at GSI. private communication.

M. Hildebrandt. Experimental investigation of β^+ - emitter yields induced by proton beams in different materials for improved modeling of in-vivo PET verification. Master's thesis, University of Heidelberg, 2012.

IAEA. Technical reports series No. 398: Absorbed dose determination in external beam radiotherapy: an international code of practice for dosimetry based on standards of absorbed dose to water. Technical report, International Atomic Energy Agency, 2000.

ICRU. Report 59: Clinical proton dosimetry - part I: Beam production, beam delivery and measurement of absorbed dose. Technical report, International Commission on Radiation Units and Measurements, 1998.

ICRU. Report 59: Nuclear Data for Neutron and Proton Radiotherapy and for Radiation Protection. Technical report, International Commission on Radiation Units and Measurements, 2000.

A. Iljinov, V. Semenov, M. Semenova, N. Sobolevsky, and L. Udovenko. Production of Radionuclides at Intermediate Energies. In *Landolt-Boernstein: Group 1*, volume vol 13a. Springer, 1991.

Y. Jongen. Commercial Ion Beam Therapy Systems. In U. Linz, editor, *Ion Beam Therapy: Fundamentals, Technology, Clinical Applications*, chapter 22. Springer, 2011. ISBN 9783642214141.

B. Kopp. Implementation of a Spot Scanning Coordinate System in a Monte Carlo Simulation for Quality Assurance at a Proton Treatment Facility, 2014.

D. Krischel. Advantages and Challenges of Superconducting Accelerators. In U. Linz, editor, *Ion Beam Therapy: Fundamentals, Technology, Clinical Applications*, chapter 23. Springer, 2011. ISBN 9783642214141.

C. Kurz, A. Mairani, and K. Parodi. First experimental-based characterization of oxygen ion beam depth dose distributions at the Heidelberg Ion-Beam Therapy Center. *Phys. Med. Biol.*, 57(15):5017–34, Aug. 2012.

H. P. Latscha, H. A. Klein, and M. Mutz. *Chemie-Basiswissen I.* Springer, 2011. ISBN 978-3-642-17522-0.

W. R. Leo. *Techniques for Nuclear and Particle Physics Experiments.* Springer, 2nd edition, 1994.

U. Linz. Physical and Biological Rationale for Using Ions in Therapy. In U. Linz, editor, *Ion Beam Therapy: Fundamentals, Technology, Clinical Applications*, chapter 4. Springer, 2011. ISBN 9783642214141.

G. MacKee. *X-rays and Radium in the Treatment of Diseases of the Skin.* Lea & Febiger, 1921.

J. Magill and J. Galy. *Radioactivity Radionuclides Radiation.* Radioactivity - Radionuclides - Radiation: Including the Universal Nuclide Chart on CD-ROM. Springer, 2004. ISBN 9783540211167.

G. Moliere. Theorie der Streuung schneller geladener Teilchen - II Mehrfachstreuung und Vielfachstreuung. *Z. Naturforsch. A Phys. Sci.*, 3(2):78–97, 1948. ISSN 0932-0784.

N. Otuka, E. Dupont, V. Semkova, B. Pritychenko, A. I. Blokhin, M. Aikawa, S. Babykina, M. Bossant, G. Chen, S. Dunaeva, R. A. Forrest, T. Fukahori, N. Furutachi, S. Ganesan, Z. Ge, O. O. Gritzay, M. Herman, S. Hlavac, K. Kato, B. Lalremruata, Y. O. Lee, A. Makinaga, K. Matsumoto, M. Mikhaylyukova, G. Pikulina, V. G. Pronyaev, A. Saxena, O. Schwerer, S. P. Simakov, N. Soppera, R. Suzuki, S. Takacs, X. Tao, S. Taova, F. Tarkanyi, V. V. Varlamov, J. Wang, S. C. Yang, V. Zerkin, and Y. Zhuang. Towards a More Complete and Accurate Experimental Nuclear Reaction Data Library (EXFOR): International Collaboration Between Nuclear Reaction Data Centres (NRDC). *NUCLEAR DATA SHEETS*, 120(SI):272–276, JUN 2014. ISSN 0090-3752.

K. Parodi. *On the feasibility of dose quantification with in-beam PET data in radiotherapy with 12C and proton beams.* PhD thesis, Technische Universität Dresden, 2004.

K. Parodi. *In Vivo Dose Verification*, chapter 16. Taylor & Francis, 2011a. ISBN 9781439836446.

K. Parodi. Monte Carlo Methods for Dose Calculations. In U. Linz, editor, *Ion Beam Therapy: Fundamentals, Technology, Clinical Applications*, chapter 7. Springer, 2011b. ISBN 9783642214141.

K. Parodi, W. Enghardt, and T. Haberer. In-beam PET measurements of beta+ radioactivity induced by proton beams. *Phys. Med. Biol.*, 47(1):21–36, Jan. 2002.

K. Parodi, F. Ponisch, and W. Enghardt. Experimental study on the feasibility of in-beam PET for accurate monitoring of proton therapy. *Nuclear Science, IEEE Transactions on*, 52(3):778–786, June 2005. ISSN 0018-9499.

K. Parodi, A. Ferrari, F. Sommerer, and H. Paganetti. Clinical CT-based calculations of dose and positron emitter distributions in proton therapy using the FLUKA Monte Carlo code. *Phys. Med. Biol.*, 52(12):3369–87, June 2007a.

K. Parodi, H. Paganetti, E. Cascio, J. B. Flanz, A. A. Bonab, N. M. Alpert, K. Lohmann, and T. Bortfeld. PET/CT imaging for treatment verification after proton therapy: a study with plastic phantoms and metallic implants. *Med Phys*, 34(2):419–35, Feb. 2007b.

K. Parodi, T. Bortfeld, and T. Haberer. Comparison between in-beam and offline positron emission tomography imaging of proton and carbon ion therapeutic irradiation at synchrotron- and cyclotron-based facilities. *Int. J. Radiat. Oncol. Biol. Phys.*, 71(3): 945–56, July 2008.

K. Parodi, A. Mairani, and F. Sommerer. Monte Carlo-based parametrization of the lateral dose spread for clinical treatment planning of scanned proton and carbon ion beams. *J. Radiat. Res.*, 54 Suppl 1:i91–6, July 2013.

K. Parodi, H. Paganetti, H. A. Shih, S. Michaud, J. S. Loeffler, T. F. DeLaney, N. J. Liebsch, J. E. Munzenrider, A. J. Fischman, A. Knopf, and T. Bortfeld. Patient study of in vivo verification of beam delivery and range, using positron emission tomography and computed tomography imaging after proton therapy. *Int. J. Radiat. Oncol. Biol. Phys.*, 68(3):920–934, JUL 1 2007. ISSN 0360-3016.

E. Pedroni, S. Scheib, T. Böhringer, A. Coray, M. Grossmann, S. Lin, and A. Lomax. Experimental characterization and physical modelling of the dose distribution of scanned proton pencil beams. *Phys. Med. Biol.*, 50(3):541–61, Feb. 2005.

W. Röntgen. *Eine neue Art von Strahlen.* Verlag und Druck der Stahel'schen K. Hof- und Universitäts-Buch- und Kunsthandlung, 1896.

Saha, G.B. *Basics of PET Imaging: Physics, Chemistry, and Regulations.* Springer, 2nd edition, 2010. ISBN 9781441908056.

G. O. Sawakuchi, U. Titt, D. Mirkovic, G. Ciangaru, X. R. Zhu, N. Sahoo, M. T. Gillin, and R. Mohan. Monte Carlo investigation of the low-dose envelope from scanned proton pencil beams. *Phys. Med. Biol.*, 55(3):711–21, Feb. 2010a.

G. O. Sawakuchi, X. R. Zhu, F. Poenisch, K. Suzuki, G. Ciangaru, U. Titt, A. Anand, R. Mohan, M. T. Gillin, and N. Sahoo. Experimental characterization of the low-dose envelope of spot scanning proton beams. *Phys. Med. Biol.*, 55(12):3467–78, June 2010b.

D. Schardt, T. Elsaesser, and D. Schulz-Ertner. Heavy-ion tumor therapy: Physical and radiobiological benefits. *Rev. Mod. Phys.*, 82(1):383–425, JAN-MAR 2010. ISSN 0034-6861.

J. Schwaab, S. Brons, J. Fieres, and K. Parodi. Experimental characterization of lateral profiles of scanned proton and carbon ion pencil beams for improved beam models in ion therapy treatment planning. *Phys. Med. Biol.*, 56(24):7813–27, Dec. 2011.

E. Seravalli, C. Robert, J. Bauer, F. Stichelbaut, C. Kurz, J. Smeets, C. V. N. Ty, D. R. Schaart, I. Buvat, K. Parodi, and F. Verhaegen. Monte Carlo calculations of positron emitter yields in proton radiotherapy. *Phys. Med. Biol.*, 57(6):1659–73, Mar. 2012.

F. Sommerer, F. Cerutti, K. Parodi, A. Ferrari, W. Enghardt, and H. Aiginger. In-beam PET monitoring of mono-energetic (16)O and (12)C beams: experiments and FLUKA simulations for homogeneous targets. *Phys. Med. Biol.*, 54(13):3979–96, July 2009.

E. Testa, M. Bajard, M. Chevallier, D. Dauvergne, F. Le Foulher, N. Freud, J. M. Letang, J. C. Poizat, C. Ray, and M. Testa. Dose profile monitoring with carbon ions by means of prompt-gamma measurements. *Nucl. Instr. Meth. Phys. Res. B*, 267(6):993–996, MAR 2009. ISSN 0168-583X. 7thTriennial International Symposium on Swift Heavy Ions in Matter, Lyon, FRANCE, JUN 02-05, 2008.

V. Vlachoudis. FLAIR: A Powerful But User Friendly Graphical Interface For FLUKA. In *Proc. Int. Conf. on Mathematics, Computational Methods & Reactor Physics (M&C 2009)*, 2009.

M. Weick-Kleemann. Dosimetrische Charakterisierung der Spotshape für zwei verschiedene Beamlines der selben Protonentherapieanlage, 2013.

R. Wilson. Radiological Use of Fast Protons. *Radiology*, 47(5):487–491, 1946. ISSN 0033-8419.

T. I. Yock and N. J. Tarbell. Technology insight: Proton beam radiotherapy for treatment in pediatric brain tumors. *Nat. Clin. Pract. Oncol.*, 1(2):97–103; quiz 1 p following 111, Dec 2004.

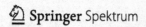

Printed in the United States
By Bookmasters